Roberta Kestenbaum

STUDY GUIDE

HUMAN DEVELOPMENT

Second Edition

F. Philip Rice

PRENTICE HALL
Englewood Cliffs, New Jersey 07632

©1995 by PRENTICE-HALL, INC.
A Simon and Schuster Company
Englewood Cliffs, New Jersey 07632

All rights reserved

10 9 8 7 6 5 4 3 2

ISBN 0-02-363155-4
Printed in the United States of America

TABLE OF CONTENTS

Part One: The Study of Human Development Over the Lifespan

Chapter 1: A Life-Span Developmental Perspective.................... 1
Chapter 2: Theories of Development.. 14

Part Two: The Beginnings of Human Life

Chapter 3: Heredity, Environmental Influences and Prenatal Development.. 31
Chapter 4: Childbirth and the Neonate................................... 48

Part Three: Child Development

Chapter 5: Perspectives on Child Development........................ 62
Chapter 6: Physical Development... 69
Chapter 7: Cognitive Development... 82
Chapter 8: Emotional Development.. 101
Chapter 9: Social Development... 114

Part Four: Adolescent Development

Chapter 10: Perspectives on Adolescent Development................ 132
Chapter 11: Physical Development.. 141
Chapter 12: Cognitive Development.. 155
Chapter 13: Emotional Development....................................... 165
Chapter 14: Social Development.. 176

Part 5: Adult Development

Chapter 15: Perspectives on Adult Development....................... 188
Chapter 16: Physical Development.. 199
Chapter 17: Cognitive Development.. 218
Chapter 18: Emotional Development....................................... 233
Chapter 19: Social Development.. 247
Chapter 20: Death, Dying and Bereavement............................ 265

How To Use This Study Guide

This study guide was prepared for you as an aid in studying and learning from your textbook. It provides chapter summaries, study questions, opportunities for you to define and use key terms, and self-test multiple choice questions that cover the material in each chapter. If you use this guide in conjunction with your textbook, you should obtain a thorough understanding of the material.

We suggest the following steps as an effective means of studying:

1. For each chapter in the textbook, first preview the chapter by glancing through at the section headings and sub-headings, and at figures, tables, and illustrations. This will begin to familiarize you with what you will be reading, and prepare you for what's ahead of you. At this point, it might be helpful to look at the **Chapter Outline** in the Study Guide and at the **Learning Objectives**. This will give you an idea of what the major points and areas of discussion are in the chapter, and should help you to focus on the important parts as you read.

2. After previewing, you should thoroughly read each chapter, keeping in mind what the learning objectives and questions are. You may want to keep the study guide open to the **Learning Objectives/Study Questions** and fill in the answers as you go along. In this way, you can be sure that you have read and understood each major point.

3. After reading each chapter, it is a good idea to sit back and reflect about what you have just read. You may want to think about the material in relation to your own childhood or upbringing, and think about whether the author's presentation of the material agrees with what you would expect. Were there issues that surprised you or clarified matters for you? Does this material fit in with what you've learned in other classes or through the media? The **Discussion Questions** at the end of each chapter in the textbook can help you to think about the material that you have just read.

4. The next step is to review the material using the study guide. You should go over the answers that you have already given for the **Study Questions** and test yourself on the material. Once you feel comfortable with these answers, go on to the next section, the **Key Terms**. For each set of key terms, give a definition for each term as best you can (you may want to use a pencil). After you have given it your best shot, go back to the textbook and check your answers, changing them in the study guide when necessary. If you use this method, rather than just initially looking it up in the text, you are more likely to learn the concepts at a deeper level. Once you feel comfortable with the definitions, go on to the **Applications** section. These sections were designed to give you the opportunity to use the key terms. Some of these fill-in-the-blank statements reiterate the definitions while others are more conceptual and/or more applied. You can fill in the blanks in the study guide and then check your answers at the end of each chapter. After completing these exercises, you should have an excellent understanding of the key terms and how they

are used.

5. The final way to test yourself on your knowledge of the material in each chapter is to do the **Self-Test Multiple Choice Questions**. These questions cover the material from the entire chapter. Some are definitional, some are factual, and some are conceptual. Thus, some of the answers will come straight from the text itself, and others will require more processing, perhaps with applications to hypothetical scenarios. You should check your answers with the answers given at the end of each chapter. If you are unclear about any of the answers, then go back to the text and read about it again (the page numbers for where the questions were derived from are provided with the correct answer choices).

After following these steps, you should have a thorough understanding of the material presented in each chapter. This should help you later on when you need to review the chapters for exams or assignments. Enjoy the material!

Chapter 1

A LIFE-SPAN DEVELOPMENTAL PERSPECTIVE

CHAPTER OUTLINE

I. Introduction to the Study of Human Development

    In this course, we seek to describe, explain, predict and influence the changes that take place in our lives.

II. Periods of Development

    A. Child Development
        1. Prenatal Period: Conception through Birth
        2. Infancy: The First Two Years
        3. Early Childhood: 3 to 5 Years
        4. Middle Childhood: 6 to 11 Years

    B. Adolescent Development
        1. Early Adolescence: 12 to 14 Years
        2. Middle or Late Adolescence: 15 to 19 Years

    C. Adult Development
        1. Early Adulthood: 20s and 30s
        2. Middle Adulthood: 40s and 50s
        3. Late Adulthood: 60 and Over

III. A Philosophy of Life-Span Development

    A. Development is multidimensional and interdisciplinary. Development can be divided into 4 dimensions: biological, cognitive, emotional, and social development.

    B. Development continues throughout the lifespan.

    C. Both heredity (nature) and environment (nurture) influence development.

D. Development reflects both continuity (a gradual, continuous process of growth and change) and discontinuity (a series of distinct stages, separated by abrupt changes).

E. Development is cumulative. Early experience can affect later development.

F. Development reflects both stability and change.

G. Development is variable. Growth is uneven, such that not all dimensions grow at the same rate.

H. Development is sometimes cyclical and repetitive.

I. Development reflects individual differences.

J. Cultural differences can affect development.

K. Developmental influences are reciprocal. Not only does the environment affect development, but children affect their environment.

IV. Research in Human Development: The Scientific Method involves:

A. Formulating the problem or the question.

B. Developing a hypothesis.

C. Testing the hypothesis.

D. Drawing conclusions.

V. Data Collection Methods and Considerations

A. Naturalistic observation involves watching people in natural environments.

B. Interviews are conducted face-to-face.

C. Questionnaires are usually completed anonymously.

D. Case studies involve longitudinal investigations of individuals rather than groups.

E. Standardized testing measures specific characteristics. Tests should have high validity and reliability.

VI. Samples should be representative of the general population.

VII. Experimental Methods

    A. Procedure - In an experiment, the variables are manipulated to determine how one affects the other.

    B. Independent and Dependent Variables - The independent variable is manipulated by the experimenter, and the dependent variable is acted upon.

    C. Establishing Relationships - Experimenters sometimes look at the relationship, or correlation, between variables.

VIII. Research Designs

    A. Age, Cohort (a group of people born during the same time period) and the Time of Testing can all affect research findings.

    B. Types of Research Designs:
       1. Cross-sectional Studies - Different age groups or cohorts are compared at one time of testing.
       2. Longitudinal Studies - One group of people is studied repeatedly over a period of time.
       3. Time Lag Studies - Different cohorts are examined using intervals in time measurements.
       4. Sequential Studies - The cross-sequential, time-sequential, and cohort-sequential designs combine aspects of the other designs.

    C. Each type of study design has advantages and disadvantages.

    D. We have learned about diverse cultures through cross-cultural research. It is important to include information from cross-cultural research when considering theories of development.

    E. Ethical issues in research - Informed consent and protection from harm are fundamental principles of ethical standards in developmental research. Parents, guardians and researchers are responsible for the welfare of the children.

LEARNING OBJECTIVES/ STUDY QUESTIONS

After reading this chapter, you should be able to:

1. Discuss what researchers of human development seek to do.

2. Describe the three major developmental periods and their subdivisions:
   a.

   b.

   c.

3. Discuss some of the different aspects of the philosophy of life-span development.
   a. Define what continuous vs. discontinuous development is.

   b. Discuss the influences of both heredity and the environment on development.

   c. Discuss what it means that development is cumulative, variable, cyclical, reciprocal, and involves both stability and change.

   d. Discuss individual differences and cultural differences.

4. Describe four different types of data collection.
   a.

   b.

   c.

   d.

5. Define what independent and dependent variables are.

6. Describe what correlational studies are.

7. Describe the four major research designs and what the advantages and disadvantages of each are.
   a.

   b.

   c.

   d.

8. Discuss the benefits of conducting cross-cultural research.

9. Discuss some of the ethical issues involved in developmental research.

KEY TERMS I

In your own words, provide a definition for each of the following terms:

1. Prenatal Period_____

2. Infancy_____

3. Early Childhood_____

4. Middle Childhood_____

5. Adolescence_____

6. Early Adulthood_____

7. Middle Adulthood _____

8. Late Adulthood _____

9. Continuity _____

10. Discontinuity _____

11. Nature _____

12. Nurture _____

13. Scientific method _____

14. Experimental methods _____

15. Naturalistic observations _____

16. Interviews _____

17. Questionnaires _____

18. Case studies _____

APPLICATIONS I

For each of the following, fill in the blank with one of the terms listed above.

1. The period of transition between childhood and adulthood during which sexual maturation takes place is _____.

2. A researcher believes that the ability to control one's own emotions is a skill that is learned from other people. This researcher emphasizes _____ over _____ as the major influence on the development of this skill.

3. A researcher follows one girl from the time she is 6 months old to when she is 5 years old, and keeps extensive records on this child's language development. This is an example of a _____.

4. The human organism develops the basic body structure and the major organs during the _____ period.

5. A person experiencing a mid-life crisis is most likely in _____.

6. A person who is very verbal would be a better candidate for the method of _____ than would a person who either can't or doesn't like to talk.

7. A developmental psychologist who believes that humans develop through a series of stages which are dramatically different from each other believes that development is _____.

8. The ability to read and write first becomes important during _____.

9. A child who is beginning to develop a sense of self and a gender identity is most likely in the age period of _____.

10. A researcher watches children playing in a schoolyard and records how often they engage in aggressive behavior. This researcher is using the method of _____.

KEY TERMS II

In your own words, provide a definition for the following terms:

1. Standardized tests _____

2. Validity _____

3. Reliability _____

4. Correlation _____

5. Random sample _____

6. Representative sample _____

7. Independent variable _____

8. Dependent variable _____

9. Cohort _____

10. Cross-sectional study _____

11. Longitudinal study _____

12. Time lag study_____
_____
13. Sequential study_____
_____
14. Cohort-sequential study_____
_____

APPLICATIONS II

For each of the following, fill in the blanks with one of the terms listed above.

1. A researcher who statistically analyzes the data to see the association between two variables is interested in the _____ between the variables.

2. Two children who were both born in 1985 are said to be of the same _____.

3. A researcher who wants to separate the effects of different ages and cohorts would be best off using a _____ design.

4. A researcher used a sample of people in her study that included the same percentages of the general population of people from different backgrounds. This sample is a _____ sample.

5. A study in which a group of 10 year olds were tested, and then two years later, another group of 10 year olds were tested is an example of a _____ study.

6. A test that was designed to measure shyness turned out to be a better measure of verbal ability. This measure would be said to have low _____ in terms of shyness.

7. A study in which a group of 3 year olds and a group of 5 year olds were compared at the same time for their ability to solve complex problems is a _____ study.

8. A researcher showed one group of children a television program with a lot of violence in it, and another group saw a pleasant show about cats and dogs. The researcher then measured the frequency of aggressive acts that were committed by the two groups. The type of television show is the _____ variable, and the measure of aggression is the _____ variable.

9. A _____ study confounds age and time of testing.

10. A group of adolescents took an intelligence test in March. When they took the test again one month later, their scores were very similar to the first time they took it. This test could be said to have high _____.

SELF-TEST MULTIPLE CHOICE QUESTIONS

Circle the best answer for each question.

1. The period that includes development from conception to birth is
   a. the prenatal period.
   b. infancy.
   c. the postnatal period.
   d. the perinatal period.

2. A child who is just beginning to use language and to form attachments to caregivers is most likely in which developmental period?
   a. prenatal
   b. infancy
   c. early childhood
   d. middle childhood

3. Which of the following is an important development during adolescence?
   a. sexual maturation
   b. the beginning of formal operational thinking
   c. becoming more independent from their parents
   d. all of the above

4. Which of the following is the best statement about development across the life-span?
   a. Most aspects of development are achieved by adolescence and there is little change after that time period.
   b. While aspects of personality are generally set by early adulthood, physical growth continues through the life-span.
   c. Aspects of intelligence do not undergo changes during adulthood, but emotionally, individuals continue to develop for several more decades.
   d. Although some aspects of physical growth may stop, most other aspects of growth and adaptation continue throughout life.

5. The critical issue in the nature vs. nurture debate is
   a. how much of development is determined by heredity.
   b. how much of development is determined by the environment.
   c. how heredity and the effects of the environment interact.
   d. whether the environment or heredity is more influential.

6. A psychologist who emphasizes how individuals' environments affect how they learn will most likely believe that development is
   a. discontinuous.
   b. continuous.
   c. stage-like.
   d. repetitive.

7. A researcher who studies whether aggressive children become violent adults is interested in
   a. the role of genetics vs. the role of the environment.
   b. whether development is multidimensional.
   c. the stability of certain personality characteristics over development.
   d. the influence of early traumatic events on later development.

8. Mary has a quick temper and she often fights with her sisters. Her sisters have come to expect that Mary will react hostilely, and so they often don't include her in their activities. This makes Mary even more angry and she fights with them even more. This little scenario demonstrates that
   a. developmental influences can act reciprocally.
   b. development is continuous.
   c. development is influenced primarily by genetics.
   d. development is not stable.

9. A researcher who unobtrusively watches fathers interact with their children on a playground while recording what types of encouragement they give to their children is engaging in which type of data collection?
   a. longitudinal
   b. cross-sectional
   c. naturalistic observation
   d. case study

10. Which of the following is not a disadvantage of using interviews as a data collection method?
    a. Interviews are very time consuming.
    b. Personal information about individuals cannot be obtained through interviews.
    c. Interviewers need extensive training in order to be non-biased.
    d. Some people tend to be less verbal than others.

11. Which of the following is an advantage of using questionnaires rather than interviews?
    a. Usually more people can be surveyed with a questionnaire than with interviews.
    b. People tend to be more honest with questionnaires than in interviews.
    c. Data is more standardized and easier to interpret with questionnaires.
    d. all of the above

12. A clinician who is interested in obtaining as much information as possible about a patient with unusual symptoms would be best off using which method of data collection?
    a. questionnaires
    b. interviews
    c. a case study
    d. standardized testing

13. A researcher designed a test for measuring levels of depression, and she wanted to find out if indeed the test really measured individuals' experiences of depression. This researcher is interested in the _____ of the measure.
    a. reliability
    b. validity
    c. sampling
    d. representativeness

14. A researcher who is interested in the reliability of a measure would be most likely to
    a. administer the test more than once to the same individuals.
    b. have at least two different examiners administer the test.
    c. both of the above
    d. neither of the above

15. If a sample is not representative of the population, a researcher cannot
    a. draw conclusions from their data about the general population.
    b. draw conclusions about the group that has been studied.
    c. apply their findings to other groups that are similar to those that have been studied.
    d. use this data to predict later developmental trends.

16. Cause and effect is most easily established in
    a. case studies.
    b. correlational studies.
    c. experimental methods.
    d. non-experimental methods.

17. A researcher is interested in the effects of age on reading skills. In this example, age would be considered the
    a. dependent variable.
    b. independent variable.
    c. correlational variable.
    d. random variable.

18. A researcher finds a statistical correlation of -.86 between how much television children watch and children's willingness to share. This means that
    a. the more television that children watch, the less likely they are to share.
    b. the more television that children watch, the more likely they are to share.
    c. how much television children watch is not related to their sharing behavior.
    d. watching a lot of television causes children to be less willing to share.

19. Which of the following is an example of a cohort?
    a. a group of children of different ages playing together at a playground
    b. a group of children of different ages who were all tested at the same time
    c. a group of siblings who were all tested at the same time
    d. a group of children who were all born in 1982

20. A group of 20-year-olds, 40-year-olds and 60-year-olds were all tested during the same time period on tests of spatial abilities. This is an example of which kind of design?
    a. longitudinal
    b. time lag
    c. cross-sectional
    d. sequential

21. In a longitudinal study, a group of 10-year-olds would be
    a. compared to a group of 12-year-olds and a group of 14-year-olds.
    b. followed for several years and tested at different intervals.
    c. compared to another group of 10-year-olds tested two years later.
    d. compared to groups of children older and younger.

22. One problem with longitudinal designs is that
    a. there may be cohort effects.
    b. there may be selective subject dropout.
    c. subjects are seen at only one time period.
    d. they occur over too short a period of time.

23. If a researcher wanted to eliminate age effects, s/he should conduct a
    a. cross-sectional study.
    b. longitudinal study.
    c. time lag study.
    d. sequential study.

24. Which of the following effects is most likely to be unimportant?
    a. age effects
    b. time of testing effects
    c. cohort effects
    d. all of the above

25. Which of the following considerations are ethically important in the study of human development?
    a. protection from harm
    b. confidentiality
    c. minimizing physical and psychological stress
    d. all of the above

ANSWER KEY

APPLICATIONS I

1. adolescence
2. nurture; nature
3. case study
4. prenatal
5. middle adulthood
6. interviews
7. discontinuous
8. middle childhood
9. early childhood
10. naturalistic observations

APPLICATIONS II

1. correlation
2. cohort
3. cohort-sequential
4. representative
5. time lag
6. validity
7. cross-sectional
8. independent; dependent
9. longitudinal
10. reliability

MULTIPLE CHOICE

1. a (p. 6)
2. b (p. 6)
3. d (p. 7)
4. d (p. 10)
5. c (p. 10)
6. b (p. 11)
7. c (p. 12)
8. a (p. 16)
9. c (p. 16)
10. b (p. 17)
11. d (p. 17)
12. c (p. 17)
13. b (p. 17)
14. c (p. 17)
15. a (p. 18)
16. c (p. 18)
17. b (p. 18)
18. a (p. 18)
19. d (p. 20)
20. c (p. 20)
21. b (p. 20)
22. b (p. 21)
23. c (p. 22)
24. b (p. 22)
25. d (p. 23)

Chapter 2

THEORIES OF DEVELOPMENT

CHAPTER OUTLINE

I. The Role of Theories

A theory organizes data, ideas, and hypotheses and states them in coherent, interrelated, general propositions, principles, or laws, which can be used in explaining and predicting phenomena.

II. Psychoanalytic Theories

    A. Freud: Psychosexual Theory
        1. Sigmund Freud originated psychoanalytic theory which emphasizes the importance of childhood experiences and unconscious motivations in influencing behavior.
        2. He used the techniques of free association and dream interpretation to explore inner conflicts between sexual urges and aggressive instincts and societal expectations.
        3. Personality is composed of 3 components: the id, the ego and the superego.
        4. People can relieve anxiety and conflict by using defense mechanisms.
        5. Freud described a psychosexual theory of development, in which the center of sensual sensitivity shifts from one body zone to another over development. The five stages are:
           a. Oral stage - centers around the mouth.
           b. Anal Stage - centers around the anus.
           c. Phallic Stage - centers around the genitals and same-sex parents; the Oedipal complex and the Electra Complex develop at this time.
           d. Latency Stage - sexual urges are repressed.
           e. Genital Stage - sexual interest in the opposite sex begins.

    B. Erikson: Psychosocial Theory

        1. Erikson disagreed with Freud on several points:
           a. Freud placed too much emphasis on the sexual basis for behavior.

       b. Freud did not place enough emphasis on adult development.
       c. Freud was too cynical.
    2. Erikson described eight stages and a psychosocial task that had to be mastered at each stage:
       a. Trust vs. distrust (0 - 1 year)
       b. Autonomy vs. shame and doubt (1 - 2 years)
       c. Initiative vs. guilt (3 - 5 years)
       d. Industry vs. inferiority (6 - 11 years)
       e. Identity vs. role confusion (12 - 19 years)
       f. Intimacy vs. isolation (young adulthood)
       g. Generativity vs. stagnation (middle adulthood)
       h. Integrity vs. despair (late adulthood)

  C. Evaluation of Psychoanalytic Theories
    1. Freud's theory helped psychotherapists gain insight into problems and helped to emphasize the importance of early experience, but he overemphasized sexual motivations and had a very cynical view of human nature.
    2. Erikson's theory is broader and encompasses the entire life-span, but he too had an antifemale bias.

III. Learning Theories

  A. Behaviorism
    1. Behaviorism emphasizes the role of environmental influences in shaping behavior.
    2. The process of learning is called conditioning.

  B. Pavlov: Classical conditioning - involves learning to associate stimuli that were not previously associated.

  C. Skinner: Operant conditioning - learning from the consequences of behavior, such as rewards and punishment.

  D. Bandura: Social Learning Theory - emphasizes the role of both cognition and environmental influences; children learn by observing others and modeling their behavior. Racial prejudice can be learned through modeling and imitation.

  E. Evaluation of Learning Theories - On the positive side, they emphasized the role of environmental influences in shaping development; however, they ignored the role of the unconscious and underlying emotions, as well as biology and maturation.

IV. Humanistic Theories - Humanists have a positive view of human nature and believe that humans should be able to reach their full potential as self-actualized persons.

A. Buhler: Developmental Phase Theory - Buhler believed that the goal of human beings is fulfillment and self-actualization, which can be reached by accomplishments.

B. Maslow: Hierarchy of Needs Theory - Maslow believed that human behavior can be explained as motivation to satisfy needs, such as physiological, safety, love and belongingness, esteem and self-actualization needs.

C. Rogers: Personal Growth Theory - Rogers believed that if people are given freedom, emotional support and unconditioned positive regard, they will develop into fully functioning human beings.

D. Evaluation of Humanistic Theories - Humanists teach people to believe in themselves and assume responsibility for their actions; however, they have been criticized for being too optimistic.

V. Cognitive Theories - Cognition is the act or process of knowing.

    A. Piaget: Cognitive Development
      1. Children learn to *adapt* through the processes of assimilation and accommodation.
      2. Piaget described four stages of cognitive development:
        a. Sensorimotor stage (0 - 2 years)
        b. Preoperational stage (2 - 7 years)
        c. Concrete operational stage (7 - 11 years)
        d. Formal operational stage (11 years and up)

    B. Information Processing - emphasizes the progressive steps, action, and operations that take place when the person receives, perceives, remembers, thinks about, and utilizes information.

    C. Evaluation of Cognitive Theories
      1. Piaget revolutionized developmental psychology by focusing on mental processes and their influence on behavior. He has been criticized for underestimating the role of the school and home in influencing behavior, and for lack of evidence of discrete developmental stages.
      2. Both Piaget and information-processing approaches ignore the role of unconscious emotions as causes of behavior.

VI. Ethological Theories

    A. Lorenz: Imprinting
      1. Ethology emphasizes that behavior is a product of evolution and is biologically determined.
      2. Lorenz described the process of imprinting, which involves rapidly

developing an attachment for the first object seen.

    B. Bonding and Attachment Theory
        1. There is some evidence to suggest that parent-infant contacts during the early hours and days of life influence later relationships, although it is not clear that there is a critical period for bonding.
        2. Bowlby suggests that over the first year of life, infants develop attachments to important individuals.

    C. Hinde: Sensitive Periods of Development - Hinde prefers the term sensitive period which is a broader and more flexible term than critical period.

    D. Evaluation of Ethological Theories - The principle of sensitive periods is useful, but too much emphasis is placed on biology.

VII. An Eclectic Theoretical Orientation - This book does not emphasize one theory, but believes that each theory contributes something to our understanding of human development.

LEARNING OBJECTIVES/ STUDY QUESTIONS

After reading this chapter, you should be able to:

1. Define what a theory is.

2. Describe the basic ideas of Freud's psychoanalytic theory.

3. Describe the three components of personality according to Freud:
   a.

   b.

   c.

4. Define the following defense mechanisms:
   a. Repression -

   b. Regression -

   c. Sublimation -

d. Displacement -

e. Reaction formation -

f. Denial -

g. Rationalization -

5. Describe the five stages of Freud's psychosexual stages, including the changing erogenous zones:
   a.

   b.

   c.

   d.

   e.

6. Discuss the points that Freud and Erikson disagreed on.

7. Describe Erikson's eight stages of psychosocial development:
   a.

   b.

   c.

   d.

   e.

   f.

   g.

h.

8. Discuss the positive and negative aspects of psychoanalytic theories.

9. Describe what classical and operant conditioning are:
   a. classical conditioning -

   b. operant conditioning -

10. Describe social learning theory.

11. Discuss the positive and negative aspects of learning theories.

12. Describe the humanistic theories of Buhler, Maslow, and Rogers:
   a. Buhler -

   b. Maslow -

   c. Rogers -

13. Describe the processes of assimilation and accommodation:
   a. assimilation -

   b. accommodation -

14. Describe Piaget's four stages of cognitive development:
   a.

   b.

c.

d.

15. Describe the information-processing approach.

16. Discuss the advantages and disadvantages of the cognitive theories.

17. Discuss the concepts of imprinting, bonding, and attachment in terms of critical and sensitive periods.

KEY TERMS I

In your own words, provide a definition for each of the following terms:

1. Theory _____

2. Psychoanalytic theory _____

3. Free association _____

4. Pleasure principle _____

5. Id _____

6. Ego _____

7. Superego _____

8. Defense mechanisms _____

9. Repression_____

10. Sublimation_____

11. Reaction formation_____

12. Denial_____

13. Rationalization_____

14. Psychosexual theory_____

15. Oral stage_____

16. Anal stage_____

17. Phallic stage_____

18. Oedipal complex_____

19. Electra complex_____

20. Latency stage_____

21. Genital stage_____

22. Fixated_____

23. Psychosocial theory_____

24. Trust vs. distrust_____

25. Autonomy vs. shame and doubt_____

26. Initiative vs. guilt_____

27. Industry vs. inferiority_____

28. Identity vs. role confusion_____

29. Intimacy vs. isolation_____

30. Generativity vs. stagnation_____

31. Integrity vs. despair _____
_____

APPLICATIONS I

For each of the following, fill in the blank with one of the terms listed above.

1. A child who hits his younger brother rather than fighting back against a boy much bigger than himself may be utilizing the defense mechanism of _____.

2. The psychosexual stage in which sexual urges are repressed is the _____.

3. The psychosexual stage in which the chief source of sensual gratification centers on the mouth is the _____ stage.

4. The psychosocial stage in which infants learn that they can trust their caregivers is _____.

5. Having a patient lie down on a couch and talk about whatever comes to mind is called the method of _____.

6. The part of the personality that operates according to the reality principle is the _____.

7. The psychosocial stage of Erikson's that is comparable to Freud's anal stage is _____.

8. A woman who continues to set a place at the table for her recently deceased husband may be engaging in the defense mechanism of _____.

9. A child who is very over-controlled and will not initiate any spontaneous behavior probably has a very highly developed _____.

10. A retired man who looks back on his life and is disappointed by what he has accomplished is in Erikson's stage of _____.

11. A boy who falls in love with his mother and is jealous of his father is experiencing an _____.

12. A man who tells his friends that he thinks that cheating on your spouse is immoral, when he himself has been attracted to other women besides his wife, may be using the defense mechanism of _____.

13. A student who claims that she failed a test because the teacher doesn't like her, when in fact she hadn't studied, may be engaging in the defense mechanism of _____.

14. An adolescent who is trying to figure out what she wants to do with her life would be in Erikson's stage of _____.

15. A third grader who feels badly about himself because he doesn't think that he is as smart as his friends has not successfully resolved Erikson's stage of _____.

KEY TERMS II

In your own words, provide a definition for each of the following terms:

1. Behaviorism _____

2. Mechanistic or deterministic _____

3. Conditioning _____

4. Classical conditioning _____

5. Operant conditioning _____

6. Positive reinforcement _____

7. Social learning theory _____

8. Modeling _____

9. Vicarious reinforcement _____

10. Vicarious punishment _____

11. Humanistic theory _____

12. Holistic view _____

13. Self-actualization _____

14. Client-centered therapy _____

15. Conditional positive regard _____

16. Unconditional positive regard _____

17. Cognition _____

18. Adaptation _____

19. Assimilation _____

20. Accommodation _____

21. Schema _____

22. Equilibration _____

23. Sensorimotor stage _____

24. Preoperational stage _____

25. Concrete operational stage _____

26. Formal operational stage _____

27. Information-processing approach _____

28. Ethology _____

29. Imprinting _____

30. Bonding _____

31. Attachment theory _____

32. Sensitive period _____

APPLICATIONS II

For each of the following, fill in the blank with one of the terms listed above.

1. A child who thinks that all blocks are made of wood is given a set of plastic blocks. According to Piaget, this child will take in this new information by the process of _____.

2. After dinner, Jane's father would turn on the dishwasher and then get Jane a bottle. Jane learned to associate the sound of the dishwasher running with getting a bottle, so that whenever the dishwasher was turned on, she began to cry for a bottle. This is an example of _____ conditioning.

3. The instinct of ducklings to follow the first moving object that they see when they are born is called _____.

4. According to Maslow, the culmination of life is achieving _____.

5. A child learned not to hit the dog after having seen his older sister get in trouble for hitting the dog. This child learned not to hit the dog based on _____ reinforcement.

6. Some researchers believe that if a mother and her newborn baby are able to spend close, physical contact within the first few days of life, this will benefit the process of _____.

7. A scientist who studies the progressive steps involved in the process of recognizing peoples' faces may be using an _____ approach.

8. A person who can use deductive and inductive reasoning would be in Piaget's _____ stage of cognitive development.

9. A therapist who believes that people should be encouraged to grow by an accepting and understanding therapist will most likely engage in _____ therapy.

10. The act of thinking is part of _____.

11. Peter was very aggressive in his preschool class, and often grabbed toys from the other children. His teachers decided that they would ignore him whenever he grabbed a toy, but they would praise him whenever he shared a toy. After a few days, Peter started to share his toys more. This is an example of _____ conditioning.

12. It is believed that children must be exposed to language during certain developmental periods in order for them to learn language. These time periods are referred to as _____.

13. Humanists hold a _____ view of human development, in which each person is considered to be more than a collection of drives, instincts and learned experiences.

14. A child who can use symbols to represent things in the world, but who can't perform mental operations that are reversible is in Piaget's _____ stage.

15. According to Piaget, young children develop _____ for performing simple acts like shaking a rattle.

SELF-TEST MULTIPLE CHOICE QUESTIONS

Circle the best answer for each question.

1. Which of the following theories emphasizes the importance of the unconscious in determining behavior?
   a. Ethology
   b. Psychoanalytic theory
   c. Humanistic theory
   d. Cognitive theory

2. According to Freud, the desire to achieve maximum pleasure and avoid pain is called
   a. the pleasure principle.
   b. the reality principle.
   c. psychic determinism.
   d. free association.

3. The part of the personality that seeks immediate gratification is the
   a. id.
   b. ego.
   c. superego.
   d. conscience.

4. A child who acts impulsively to get what she wants and doesn't think about the consequences of her actions is responding primarily to which part of the personality structure?
   a. id
   b. ego
   c. superego
   d. conscience

5. When conscious desires are forced into the unconscious, but they still exert an influence on behavior, then _____ can be said to have occurred.
   a. regression
   b. displacement
   c. sublimation
   d. repression

6. Mary was toilet-trained by the time she was 3 years old. However, when she was 4 years old, her little brother was born. Mary began to want to wear diapers again. Which defense mechanism was Mary using?
   a. reaction formation
   b. denial
   c. regression
   d. sublimation

7. During the Phallic stage, boys experience _____ and girls experience _____.
   a. penis envy; castration anxiety
   b. the Oedipal complex; the Electra complex
   c. a sense of inferiority; a sense of superiority
   d. all of the above

8. Which of the following did Freud and Erikson agree upon?
   a. that there was a strong sexual basis for behavior
   b. that there were different developmental stages throughout the life-span
   c. that early experience affects later development
   d. none of the above

9. A child who is learning how to feed himself, and who gets upset when his mother tries to interfere is probably in which of Erikson's psychosexual stages?
   a. trust vs. distrust
   b. autonomy vs. shame and doubt
   c. initiative vs. guilt
   d. industry vs. inferiority

10. Which of the following theories contends that development is continuous rather than in a stage-like fashion?
    a. Behaviorism
    b. Freudian theory
    c. Piagetian theory
    d. Eriksonian theory

11. A child who tripped and fell over a rock when he was wearing a new pair of shoes later refused to wear the shoes again because he associated them with the pain of falling down. According to classical conditioning theory, the shoes would be considered the
    a. unconditioned stimulus.
    b. unconditioned response.
    c. conditioned response.
    d. conditioned stimulus.

12. _____ is when a person learns from the consequences of her own behavior.
    a. Classical conditioning
    b. Modeling
    c. Operant conditioning
    d. Positive reinforcement

13. According to Bandura, children can learn
    a. only from the punishment they receive for their own behavior.
    b. only from the rewards they receive for their own behavior.
    c. from associating their actions with the consequences in their environment.
    d. from watching the behavior of others.

14. A psychologist who believes that to be a fully functioning individual one should strive to realize their full potential as a self-actualized person is most likely a
    a. humanist.
    b. psychoanalytic theorist.
    c. cognitive theorist.
    d. behaviorist.

15. According to Maslow, human behavior is motivated by
    a. unconscious instincts and drives.
    b. the desire to achieve equilibrium.
    c. the desire to satisfy needs.
    d. the desire to resolve inner conflicts.

16. In Maslow's hierarchy of needs, which type of need must first be satisfied before an individual can attend to the others?
    a. love needs
    b. physiological needs
    c. esteem needs
    d. self-actualization needs

17. According to Rogers, when a person has a poor self-image, she needs
    a. conditional positive regard.
    b. unconditional positive regard.
    c. to learn to accept her real self.
    d. to learn to accept her ideal self.

18. When a child uses information that he already has and applies it to something in the environment, this child is
    a. in a state of disequilibrium.
    b. being classically conditioned.
    c. accommodating.
    d. assimilating.

19. According to Piaget, children are motivated to develop cognitively by
    a. the desire to be more mature.
    b. the desire to achieve equilibrium.
    c. the desire to satisfy needs.
    d. the desire to achieve disequilibrium.

20. A young child who is beginning to coordinate what he sees with what he touches would be in which of Piaget's stages of cognitive development?
    a. sensorimotor
    b. preoperational
    c. concrete operational
    d. formal operational

21. Which of the following tasks could a child who was in the concrete operational stage successfully perform?
    a. using algebraic symbols
    b. solving hypothetical problems
    c. using inductive reasoning
    d. sorting into hierarchical classifications

22. Which of the following is a valid criticism of Piaget's theory of cognitive development?
    a. He underestimated the role of the school in fostering cognitive development.
    b. Many people never reach the highest stages of development, thus development through the stages may not be universal.
    c. The role of unconscious emotions as causes of behavior is ignored.
    d. all of the above

23. The emotional tie between the parent and the newborn baby that is created after intimate, physical contact in the first few days of life is referred to as
    a. imprinting.
    b. bonding.
    c. attachment.
    d. sensitivity.

24. According to Bowlby, infants develop attachments
    a. in the first few days of life.
    b. to whomever they see first when they are born.
    c. first to an individual and then to people in general.
    d. first to people in general and then to specific individuals.

25. If development is affected more during certain time periods than others, then that development can be said to have
    a. a developmental lag.
    b. a reactionary period.
    c. a sensitive period.
    d. none of the above

ANSWER KEY

APPLICATIONS I

1. displacement
2. latency stage
3. oral
4. trust vs. distrust
5. free association
6. ego
7. autonomy vs. shame and doubt
8. denial
9. superego
10. integrity vs. despair
11. Oedipal complex
12. reaction formation
13. rationalization
14. identity vs. role confusion
15. industry vs. inferiority

APPLICATIONS II

1. accommodation
2. classical
3. imprinting
4. self-actualization
5. vicarious
6. bonding
7. information-processing
8. formal operational
9. client-centered
10. cognition
11. operant
12. sensitive periods
13. holistic
14. preoperational
15. schemas

MULTIPLE CHOICE

1. b (p. 30)
2. a (p. 31)
3. a (p. 31)
4. a (p. 31)
5. d (p. 31)
6. c (p. 31)
7. b (p. 32)
8. c (p. 35)
9. b (p. 35)
10. a (p. 36)
11. d (p. 36)
12. c (p. 37)
13. d (p. 38)
14. a (p. 40)
15. c (p. 41)
16. b (p. 41)
17. b (p. 43)
18. d (p. 45)
19. b (p. 45)
20. a (p. 45)
21. d (p. 45)
22. d (p. 46)
23. b (p. 48)
24. d (p. 48)
25. c (p. 49)

Chapter 3

HEREDITY, ENVIRONMENTAL INFLUENCES, AND PRENATAL DEVELOPMENT

CHAPTER OUTLINE

I. Reproduction

   A. Spermatogenesis
      1. Through repeated cell division (meiosis), about 300 million sperm are produced daily.
      2. During intercourse, sperm are ejaculated into the vagina and travel through the uterus and the fallopian tubes.

   B. Oogenesis - the process by which the female egg cells (ova) are ripened in the ovaries.

   C. Conception - The ovum is released and fertilization by the sperm takes place in the Fallopian tube.

II. Family Planning - The goal is to enable people to have the number of children they want, when they want to have them.

   A. Benefits - Some benefits include: protecting the health of the mother and children, protecting against child abuse and neglect, and helping the marriage.

   B. Contraceptive Failure - The pill is the most effective reversible method, while periodic abstinence and spermicides are more likely to fail. Failure results mostly from improper and irregular use of contraceptive methods. Condom use has increased because they reduce the risk of sexually transmitted diseases such as AIDS.

III. Prenatal Development

    A. The Germinal Period
       1. The fertilized ovum is called a zygote.
       2. About 30 hours after fertilization, the process of cell division begins.
       3. After division, the blastula attaches itself to the inner lining of the uterus.

    B. The Embryonic Period - begins the end of the 2nd week
       1. The human embryo resembles those of other vertebrate animals.
       2. The backbone and heart form.

    C. The Fetal Period - begins at around 2 months
       1. The fetus has developed the first bone structure.
       2. By the end of the first trimester, most major organs are present, the head and face are well formed, and a heartbeat can be detected.
       3. By the 5th month, the mother can usually feel fetal movement.
       4. During the third trimester, the head and body become more proportionate.

IV. Prenatal Care

    A. Medical and Health Care - Good prenatal health care is extremely important. Mortality rates are considerably higher among minority groups who may not receive adequate prenatal care.

    B. Minor Side Effects - Normal pregnancies may include: nausea, heartburn, hemorrhoids, backache, varicose veins and other minor symptoms.

    C. Major Complications - Some complications which are a more serious threat are: pernicious vomiting, toxemia, threatened abortion, placenta praevia, tubal pregnancy, and Rh incompatibility.

V. Infertility

    A. Causes - About 20% of infertility cases involve both partners; 40% involve the man; 40% involve the woman.

    B. Fertility problems can cause significant emotional and psychological distress. Some couples report greater marital discord, whereas others say that their problems brought them closer together.

    C. Alternate Means of Conception:
       1. Artificial insemination - The sperm are injected into the vagina or uterus.
       2. Surrogate mother - Another woman is inseminated with sperm of the man.
       3. In vitro fertilization - The egg is removed from the mother, fertilized in the laboratory and replaced in the mother.

    4. Gamete intra fallopian transfer - A thin tube with the sperm and egg is inserted directly into the fallopian tube.
    5. Embryo transplant - The embryo is removed from the donor and placed into the woman's uterus.

VI. Heredity

   A. Chromosomes, Genes and DNA - Chromosomes in the nucleus of each cell carry the hereditary material called genes which are made up of DNA and control the characteristics that are inherited.

   B. The 23rd Pair and Sex Discrimination - Each sperm cell and ovum contains 23 chromosomes; the 23rd pair, the sex chromosomes, determine whether the offspring will be a boy or girl.

   C. Multiple Births - Twins may be either monozygotic (identical) or dizygotic (fraternal).

   D. Simple Inheritance and Dominant-Recessive Inheritance - When an organism inherits competing traits, only one trait will be expressed; this trait is dominant over the recessive trait.

   E. Incomplete Dominance - Sometimes one trait is not completely dominant over the other.

   F. Polygenic Inheritances - Often traits result from combinations of many genes.

   G. Sex-Linked Traits - Some defective or recessive genes are carried on only the sex chromosomes and produce sex-linked disorders.

VII. Hereditary Defects

   A. Causes of Birth Defects - 1) hereditary factors, 2) faulty environments, 3) birth injuries

   B. Genetic Defects - Some defects may be caused by a single, dominant defective gene, by pairs of recessive genes, or by multiple factors.

   C. Chromosomal Abnormalities - either sex chromosomal abnormalities or autosomal chromosomal abnormalities; Down's syndrome is a common autosomal abnormality.

   D. Genetic Counseling - For couples at risk for chromosomal abnormalities, there are a number of different types of tests such as amniocentesis, sonograms, fetoscope, and chorionic villus sampling.

VIII. Prenatal Environment and Influences

   A. Teratogens - any substances that cross the placental barrier and cause harm

   B. Drugs - Medications, narcotics, sedatives, analgesics, alcohol, tranquilizers, antidepressants, nicotine, cocaine, and marijuana can all harm the fetus.

   C. Chemicals, Heavy Metals, Environmental Pollutants - Substances such as dioxin, PCB, and lead can cause birth defects.

   D. Radiation - Radiation from atomic bombing or X-rays can cause problems.

   E. Heat - Very high temperatures can harm a fetus.

   F. Maternal Diseases - Diseases such as rubella, toxoplasmosis, sexually transmitted diseases, AIDS, and others can harm the fetus.

   G. Other Maternal Factors - Other factors such as maternal age (either very young or very old), nutrition, and stress can also affect the fetus.

IX. Heredity-Environment Interaction

   A. Studying the Influence of Heredity and Environment
      1. In twin studies, identical and fraternal twins are compared.
      2. In adoption studies, adopted children are compared to their adoptive parents and their biological parents.

   B. Influences on Personality and Temperament - Heredity not only influences intelligence, but also personality factors and temperament.

   C. Some Disorders Influenced by Heredity and Environment: alcoholism, schizophrenia, depression, infantile autism

   D. Paternal Factors in Defects - Some factors which may have effects: advanced paternal age, chronic marijuana use, alcoholism, exposure to radiation and other substances.

LEARNING OBJECTIVES/ STUDY QUESTIONS

After reading this chapter, you should be able to:

1. Describe the process of conception.

2. Discuss the benefits of family planning and the pros and cons of different methods of contraception.

3. Describe the three periods of prenatal development.
   a.

   b.

   c.

4. Discuss some of the minor and major side effects or complications of pregnancy.

5. Discuss the causes and consequences of infertility, and some of the alternate means of conception.

6. Discuss how sex of the baby is determined.

7. Describe the process of dominant-recessive inheritance.

8. List some of the sex-linked traits and disorders.

9. List some of the disorders caused by:
   a. dominant genes -

   b. recessive genes -

c. multiple factors -

10. Describe some of the chromosomal abnormalities that can arise.

11. Describe some of the tests that can be used for genetic counseling.

12. Describe some of the factors in the environment which can affect the health of the developing fetus.

13. Discuss some of the maternal diseases that can cause birth defects.

14. Discuss how maternal age, nutrition, and stress level can affect the pregnancy.

15. Discuss the two principle methods of sorting out the relative influences of heredity and environment on development.
    a.

    b.

16. Discuss some of the disorders that are influenced by both heredity and environment.

17. Discuss some of the paternal factors that can cause birth defects.

KEY TERMS I

In your own words, provide a definition for each of the following terms:

1. Gametes
2. Zygote
3. Spermatogenesis
4. Meiosis
5. Oogenesis
6. Ova
7. Ovulation
8. Fertilization or conception
9. Germinal period
10. Embryonic period
11. Fetal period
12. Morula
13. Blastula

14. Blastocyst _____

15. Implantation _____

16. Ectopic pregnancy _____

17. Embryo _____

18. Fetus _____

19. Trimester _____

20. Viable _____

21. Infertile _____

22. Artificial insemination _____

23. Homologous insemination (AIH) _____

24. Heterologous insemination (AID) _____

25. Surrogate mother _____

26. Gamete intro fallopian transfer (GIFT) _____

27. In vitro fertilization _____

28. Embryo transplant _____

APPLICATIONS I

For each of the following, fill in the blank with one of the terms listed above.

1. The blastula attaches itself to the uterine wall during the _____ period.

2. The male sperm cell and the female ovum are referred to as _____.

3. The process by which female gametes are ripened in the ovaries is called _____.

4. Vanessa has her husband's sperm injected into her vagina in an attempt to become pregnant. This procedure is referred to as _____.

38

5. Monica has the sperm of a friend of hers injected into her vagina in an attempt to become pregnant. This procedure is referred to as _____.

6. When a blastocyst implants itself in the fallopian tube, the mother is said to have a(n) _____.

7. A pregnant woman can start to feel movement during the _____ period.

8. Laura and Michael want to have a child, but Laura is infertile. Kathy is then artificially inseminated with Michael's sperm, and then after five days, the embryo is removed from Kathy and placed into Laura's uterus. This procedure is called _____.

9. The fertilized ovum is called a _____.

10. During the _____ period, the heart forms and begins beating.

11. The inner layer of the blastula that develops into the embryo is called the _____.

12. The procedure of _____ involves removing an egg from the mother and fertilizing it with the father's sperm in the laboratory, then implanting it in the uterine wall.

13. When the fetus is able to live on its own, it is said to be _____.

KEY TERMS II

In your own words, provide a definition for each of the following terms.

1. Chromosomes _____

2. Genes _____

3. DNA _____

4. Autosomes _____

5. Sex chromosomes _____

6. Monozygotic twins _____

7. Dizygotic twins_____

8. Siamese twins_____

9. Law of dominant inheritance_____

10. Dominant_____

11. Recessive_____

12. Alleles_____

13. Homozygous_____

14. Heterozygous_____

15. Phenotype_____

16. Genotype_____

17. Incomplete dominance_____

18. Polygenetic system of inheritance_____

19. Reaction range_____

20. Canalization_____

21. Sex-linked disorders_____

22. Congenital deformity_____

23. Amniocentesis_____

24. Sonogram_____

25. Fetoscope_____

26. Chorionic villi sampling_____

27. Teratogen_____

APPLICATIONS II

For each of the following, fill in the blank with one of the terms listed above.

1. Twenty-two of the 23 pairs of chromosomes contained in each sperm and ovum are called _____; the 23rd pair are the _____.

2. Barry inherits one sickle-cell gene and one normal gene, and thus has a mixture of normal and sickle-cell blood cells. These alleles can be said to be _____.

3. Andy and Tom are twins, but they don't share any more genetic material than do siblings. They are _____ twins.

4. Ed and Jeff are twins that developed from the same ovum. They are _____ twins.

5. A child who is born with a defective condition such as blindness can be said to have a _____.

6. _____ are the rod-like structures in the nucleus of every cell.

7. If Donna inherited the genetic material for curly hair from her mother and the genetic material for straight hair from her father, and she has curly hair, the gene for curly hair would be said to be _____ and the gene for straight hair would be said to be _____.

8. If Donna had inherited the genes for curly hair from both of her parents, she would be _____ for that trait.

9. Donna's underlying genetic pattern of one curly-haired gene and one straight-haired gene is referred to as the _____, while the observable trait of curly hair is called the _____.

10. Barbara has inherited the possibility of having an IQ from 105 to 115. This span of possibilities is referred to as the _____.

11. The procedure in which a hollow needle is inserted into the mother's abdomen to obtain a sample of fluid containing fetal cells is called _____.

12. A _____ uses high-frequency sound waves to obtain a visual image of the fetus.

13. Substances such as alcohol, cocaine, and nicotine are called _____ and can harm a developing fetus.

SELF-TEST MULTIPLE CHOICE QUESTIONS

Circle the best answer for each question.

1. When a sperm cell fuses with an ovum, it forms a single new cell called
   a. an embryo.
   b. a zygote.
   c. a blastula.
   d. a morula.

2. About 300 million sperm are produced daily by a human male through the process of
   a. oogenesis.
   b. fertilization.
   c. ovulation.
   d. spermatogenesis.

3. Egg cells, or ova, are ripened in the ovaries through the process of
   a. oogenesis.
   b. fertilization.
   c. ovulation.
   d. spermatogenesis.

4. Fertilization of the egg normally takes place in the
   a. vagina.
   b. fallopian tube.
   c. uterus.
   d. ovary.

5. Which of the following methods of contraception is the least likely to fail?
   a. condom
   b. diaphragm
   c. periodic abstinence
   d. pill

6. Once the cells have divided into about 100-150 cells, the zygote is called the
   a. embryo.
   b. fetus.
   c. blastula.
   d. morula.

7. During the embryonic period, the embryo has
   a. developed all of the major organs.
   b. developed full arms and legs, including fingers and toes.
   c. a tail and traces of gills which will soon disappear.
   d. well-formed eyes and eyelids.

8. A pregnant woman who has high blood pressure, waterlogging of the tissues, albumin in her urine, headaches, blurry vision and convulsions most likely has
   a. toxemia.
   b. placenta praevia.
   c. teratogens.
   d. anoxia.

9. The premature separation of the placenta from the uterine wall is called
   a. ectopic pregnancy.
   b. toxemia.
   c. tubal pregnancy.
   d. placenta praevia.

10. Which of the following is the best statement on the causes of infertility.
    a. In the majority of the cases, the problem involves the man.
    b. In the majority of the cases, the problem involves the woman.
    c. The problem involves the man or the woman about equally.
    d. Usually, the problem is with both the man and the woman.

11. Which of the following couples is at greatest risk for having trouble adjusting to involuntary childlessness?
    a. a couple who has been married and childless for a long time
    b. a husband and wife who both work
    c. a husband who works and a wife who does not work
    d. a couple who seeks treatment

12. If a woman's fallopian tubes are blocked, but she wants to carry the baby herself, her best option for an alternate means of conception would probably be
    a. to have a surrogate mother.
    b. artificial insemination.
    c. in vitro fertilization.
    d. none of the above.

13. When the sperm and the ovum unite, how many chromosomes are in the resulting zygote?
    a. 2
    b. 22
    c. 23
    d. 46

14. If a zygote contains an X chromosome from the mother and an X chromosome from the father, the child will be
    a. a girl.
    b. a boy.

15. When two ova are fertilized by two separate sperm, the two children will be
    a. Siamese twins.
    b. monozygotic twins.
    c. dizygotic twins.
    d. identical twins.

16. A child receives a gene for dark hair and one for light hair, but has dark hair. His dark hair is considered the
    a. allele.
    b. incomplete gene.
    c. genotype.
    d. phenotype.

17. If the mother has brown eyes and the father has brown eyes and the gene for brown eyes is dominant, the child
    a. will have brown eyes.
    b. will have blue eyes.
    c. could have either brown eyes or blue or green eyes.

18. Bobby grew up in a deprived environment in which there was very little intellectual stimulation. In spite of this, he still had a very high IQ. This trait could be said to be
    a. environmentally influenced.
    b. dominant.
    c. canalized.
    d. polymorphous.

19. An example of a sex-linked disorder is
    a. hemophilia.
    b. Down's syndrome.
    c. Huntington's disease.
    d. Tay-Sachs disease.

20. Defects such as cleft palate and spina bifida are caused by
    a. a single, dominant defective gene.
    b. pairs of recessive genes.
    c. multiple factors.
    d. defective sex chromosomes.

21. A major disadvantage of using _____ is that it cannot be performed until the second trimester of pregnancy.
    a. amniocentesis
    b. a sonogram
    c. a fetoscope
    d. chorionic villus sampling

22. Teratogens can cause the most damage during
    a. the first two weeks after conception.
    b. the first eight weeks after conception.
    c. the 2nd trimester.
    d. the 3rd trimester.

23. Which of the following is considered to be potentially damaging to the fetus?
    a. exposure to X-rays
    b. exposure to lead
    c. exposure to gaseous anesthetics
    d. all of the above

24. Doctors warn pregnant women not to change the cat's litter because they risk getting
    a. rubella.
    b. toxoplasmosis.
    c. syphilis.
    d. chlamydial infections.

25. Twin studies to sort out environmental vs. hereditary effects are based on the premise that
    a. monozygotic twins are more similar genetically than dizygotic twins who are more similar genetically than siblings.
    b. dizygotic twins are 100% genetically identical, but monozygotic twins are only 50% genetically similar.
    c. dizygotic twins are compared to other siblings because they share the same amount of genetic material.
    d. monozygotic twins are genetically identical while dizygotic twins share the same amount of genetic material as do other siblings.

26. Which of the following is believed to be influenced by both hereditary and environmental factors?
    a. intelligence
    b. temperament
    c. alcoholism
    d. all of the above

27. Which of the following is the best statement regarding paternal factors in hereditary defects?
    a. Many factors, such as men's exposure to radiation or lead, chronic marijuana use and alcoholism, have been associated with problems.
    b. To date, research has not yet been able to show that paternal factors such as drug and alcohol use will affect reproduction in any way.
    c. Paternal use of alcohol or drugs will only affect the baby after it is born.

ANSWER KEY

APPLICATIONS I

1. germinal
2. the gametes
3. oogenesis
4. homologous insemination
5. heterologous insemination
6. ectopic pregnancy
7. fetal
8. embryo transplant
9. zygote
10. embryonic
11. blastocyst
12. in vitro fertilization
13. viable

APPLICATIONS II

1. autosomes; sex chromosomes
2. heterozygous
3. dizygotic
4. monozygotic
5. congenital deformity
6. Chromosomes
7. dominant; recessive
8. homozygous
9. genotype; phenotype
10. reaction range
11. amniocentesis
12. sonogram
13. teratogens

MULTIPLE CHOICE

1. b (p. 58)
2. d (p. 58)
3. a (p. 59)
4. b (p. 59)
5. d (p. 60)
6. c (p. 61)
7. c (p. 63)
8. a (p. 65)
9. d (p. 65)
10. c (p. 67)
11. c (p. 67)
12. c (p. 68)
13. d (p. 69)
14. a (p. 70)
15. c (p. 71)
16. d (p. 72)
17. c (p. 72)
18. c (p. 75)
19. a (p. 75)
20. c (p. 76)
21. a (p. 76)
22. b (p. 79)
23. d (p. 80)
24. b (p. 81)
25. d (p. 85)
26. d (pp. 86-88)
27. a (p. 91)

Chapter 4

CHILDBIRTH AND THE NEONATE

CHAPTER OUTLINE

I. Childbirth

   A. Prepared Childbirth - refers to the physical, social, intellectual, and emotional preparation for the birth of a baby.

   B. Methods of Natural Childbirth
      1. Dick-Read method - emphasizes education about birth, physical conditioning and exercises, and emotional support from the partner.
      2. Lamaze method - emphasizes education about birth, physical conditioning, controlled breathing and emotional support.
      3. Prepared childbirth now often includes the father throughout the entire process.

   C. Birthing Rooms and Family Centered Care
      1. Birthing rooms are informal, pleasant home-like rooms within the hospital.
      2. Birthing centers are located apart, but near, a hospital.

II. Labor

   A. Beginning
      1. Real labor is rhythmic in nature and recurs at fixed intervals.
      2. Sometimes the first sign of labor is rupture of the amniotic sac.

   B. Duration - The median number of hours for first labor is 10.6; for all subsequent labors, 6.2.

   C. Stages
      1. The dilation stage is the longest stage, during which the cervix gradually opens.
      2. The second phase involves the passage of the baby through the birth canal.
      3. The passage of the placenta or afterbirth is the final phase.

D. Use of Anesthesia - Either general anesthesia, which can affect the fetus, or local anesthesias, which have minimal effects on the baby, can be used.

E. Fetal Monitoring - External devices are sometimes employed for normal pregnancies, and internal methods for high risk or problem pregnancies.

III. Delivery

A. Normal Delivery - The head is usually delivered first.

B. Stress on the Infant - Stress of birth produces large amounts of adrenaline and noradrenaline in the baby's blood.

C. Delivery Complications - may include vaginal bleeding, abnormal fetal heart rate, disproportion in size between the fetus and pelvic opening, or abnormal fetal presentations.

D. Anoxia and Brain Injury - Two serious delivery complications are anoxia (oxygen deprivation to the brain) and brain injury.

IV. Postpartum

A. Leboyer Method - emphasizes gentle, loving treatment of the newborn, with dim lights, soothing and massage while resting on mother's abdomen.

B. Evaluating Neonatal Health and Behavior
    1. Apgar score - has designated values for various neonatal signs, and permits a tentative and rapid diagnosis of major problems.
    2. Brazelton assessment - evaluates both the neurological condition and the behavior of the neonate.

C. Postpartum Blues - After delivery, woman may suffer varying degrees of postpartal depression.

D. Cultural factors influence postpartum care and may affect how much rest and attention a woman receives.

V. Premature and Small-for-Gestational Age (SGA) Infants

A. Classifications - Each newborn may be classified as full-term, premature, or postmature.

B. Premature Infants - born before 37 weeks gestation
    1. Problems relate to immaturity of the organs.

2. When there are later developmental deficits, it's usually because of significant medical complications.

C. Small-for-Gestational Age (SGA) Infants - Weight is below the 10th percentile, whether premature, full-term or postmature.
1. Perinatal asphyxia (lack of oxygen) and hypoglycemia (low blood sugar) can be problems for these infants.
2. Some studies suggest that the lower the birthweight, the more likely the child will have problems later on.

D. Parental Roles and Reactions - The birth of a preterm infant is a stressful event, increased by lack of social supports.

VI. The Neonate

A. Physical Appearance and Characteristics - At birth, the average full-term is about 20 inches long and about 7 pounds.

B. Physiological Functioning - Newborns breathe on their own, have a strong sucking response and have difficulty maintaining stable body temperature.

C. The Senses and Perception
1. Vision - Vision is the least developed of the senses, but by six months, visual acuity is about normal; Newborns prefer looking at faces over objects.
2. Hearing - At birth, hearing is slightly less sensitive than is adults' hearing.
3. Smell - Newborns are responsive to various odors, including odors from their mother's breast.
4. Taste - Newborns can discriminate among various taste stimuli.
5. Touch and pain - There is strong evidence of touch and pain sensitivity.

D. Reflexes - Some reflexes are: rooting and sucking, Palmar's grasp, Moro reflex, blinking, and rage reflex.

E. Brain and Nervous System
1. New nerve cells continue to form until about the second month.
2. The nerve cells continue to mature.
3. Some of the neurons are not yet myelinated.
4. The newborn's brain is immature, with some areas dysfunctional.

F. Individual Differences - Even in the first few weeks of life, infants show different temperaments.

G. Sudden Infant Death Syndrome (SIDS) occurs when the baby is sleeping, and is the most common cause of death before 1 year.

LEARNING OBJECTIVES/ STUDY QUESTIONS

After reading this chapter, you should be able to:

1. Describe methods of natural childbirth.

2. Describe the 3 stages of labor.
   a.

   b.

   c.

3. Discuss some of the effects of general and local anesthesias.

4. Discuss some of the complications that can arise during delivery.

5. Discuss two of the ways to evaluate neonatal health and behavior, including what they measure.
   a.

   b.

6. Discuss some of the pros and cons of circumcision.

7. Define the following classifications:
   a. premature -

   b. full-term -

   c. postmature -

8. Discuss some of the problems that can arise when an infant is born prematurely.

9. Discuss some of the problems that can arise for small-for-gestational age infants.

10. Discuss the physical characteristics and the physiological functioning of the neonate.

11. Describe what the newborn's five senses are like after birth.
    a. Vision -

    b. Hearing -

    c. Smell -

    d. Taste -

    e. Touch -

12. Describe the newborn's reflexes.

13. Discuss how the newborn brain differs from the adult brain.

KEY TERMS I

In your own words, provide a definition for each of the following terms:

1. Prepared childbirth _____

2. Dick-Read method _____

3. Lamaze method _____

4. Labor _____

5. Show _____

6. Amniotic sac _____

7. General anesthesia _____

8. Local anesthesia _____

9. Episiotomy _____

10. Perineum _____

11. Cesarean section _____

12. Anoxia _____

13. Prolapsed cord _____

APPLICATIONS I

For each of the following, fill in the blank with one of the terms listed above.

1. A woman who discharges blood-tinged mucous shortly before labor begins has experienced the _____.

53

2. A woman who begins to experience uterine contractions at fixed intervals of about 15 minutes apart is in _____.

3. The _____ method of natural childbirth was developed by an English obstetrician because he believed that a mother's fear created tension, which interfered in a negative way with childbirth.

4. If a pregnant woman cannot deliver her baby vaginally, the doctor may perform a _____ and remove the baby surgically.

5. If the umbilical cord is squeezed between the baby's body and the birth canal, the baby may experience _____.

6. When a pregnant woman experiences a gush of watery fluid from her vagina, then her _____ has ruptured.

7. The _____ method of natural childbirth emphasizes controlled breathing as a way to reduce pain during childbirth.

8. _____ anesthesia crosses the placental barrier and can decrease the responsiveness of the newborn baby.

9. In order to prevent excessive tearing of the tissues of the _____, doctors usually perform _____ during childbirth.

10. A mother who wants to reduce the pains of labor, but still be aware of what is happening, may opt for _____ anesthesia.

KEY TERMS II

In your own words, provide a definition for each of the following terms.

1. Leboyer method _____

2. Apgar score _____

3. Brazelton Neonatal Behavior Assessment Scale _____

4. Postpartal depression _____

5. Circumcision _____

6. Full-term infant _____

7. Premature infant _____

8. Postmature infant _____

9. Small-for-gestational age infant _____

10. Asphyxia _____

11. Hypoglycemia _____

12. Vernix caseosa _____

13. Colostrum _____

14. Reflexes _____

15. Neurons _____

APPLICATIONS II

For each of the following, fill in the blank with one of the terms listed above.

1. A child is given a value called the _____ at one minute and at 5 minutes after birth.

2. When Maria was born at 38 weeks, her weight was below the 10th percentile. She would be classified as a _____ infant.

3. Since Maria's weight was so low at birth, she may have experienced _____ and/or _____.

4. A few days after bringing her baby home from being born, Sarah felt very sad, and had bouts of crying and irritability. She was probably suffering from _____.

5. The _____ serve to transmit messages from one to another in the nervous system.

6. A child who is born into an environment in which the lights are dimmed, and where she is bathed in water of similar temperature to the amniotic fluid has probably been born by the _____.

7. If a doctor wants to know how a newborn's central nervous system is functioning, she could use the _____.

8. Although some doctors believe that _____ helps prevent cancer of the penis and urinary tract infections in their partners, other doctors think that it is a procedure that is medically unnecessary.

9. Bill was born after 35 weeks. He would be classified as _____.

10. Some _____ have developed through evolution in order to protect the baby from physical discomfort or from falling.

SELF-TEST MULTIPLE CHOICE QUESTIONS

Circle the best answer for each question.

1. The person who conceived of the idea of prepared childbirth because he believed that if women were educated about their pregnancies, it would eliminate their anxieties was
    a. Dick-Read.
    b. Lamaze.
    c. Leboyer.
    d. Brazelton.

2. John and Kelly are taking classes together to learn about the birth of their child, including exercises that can be done and how to control breathing during labor. This couple probably plans to use which method of natural childbirth?
    a. the Dick-Read method
    b. the Lamaze method
    c. the Leboyer method
    d. the Brazelton method

3. Studies of nurse-midwifery practice have consistently shown delivery outcomes as good or better than physicians'. One reason for this is that
    a. it has been shown that midwives are more skilled than physicians at delivering babies.
    b. midwives have better access to medical equipment when it is necessary.
    c. midwives are better trained in emergency procedures.
    d. the prospective parents who want to use a midwife must be screened beforehand and should be in good health.

4. Discharge of the blood-tinged mucous plug that seals the neck of the uterus is called
    a. labor.
    b. show.
    c. a contraction.
    d. dilation.

5. Often the first sign of impending labor is
   a. the discharge of the placenta.
   b. contractions half a minute apart.
   c. the rupture of the amniotic sac.
   d. all of the above

6. The median number of hours of labor for pripara women is
   a. 4.2.
   b. 6.2.
   c. 10.6.
   d. 14.6.

7. Over a period of several hours, Misu's cervix is gradually opening. Misu is in which stage of labor?
   a. first
   b. second
   c. third
   d. fourth

8. The mouth of the cervix should open to about how much for a normal delivery?
   a. 2 inches
   b. 3 inches
   c. 4 inches
   d. 5 inches

9. Electronic fetal monitoring with internal leads is generally used
   a. for most routine normal pregnancies.
   b. during normal pregnancies when parents want to monitor the pregnancy at all times.
   c. when problems with the pregnancy are anticipated.
   d. when doctors want to restrict the movement of the baby.

10. Research has suggested that the reason that many babies are born alert is
    a. because of the change in temperature from the womb to air.
    b. that they are startled by the noises when they are first born.
    c. that they are affected by the bright lights in the delivery room.
    d. that the stress of birth stimulates production of adrenaline and noradrenaline.

11. In a breech presentation, the _____ present first rather than the head.
    a. buttocks
    b. face
    c. brow
    d. shoulder

12. A neonate who does not get enough oxygen to the brain during the birth process can be said to have experienced
    a. toxemia.
    b. anoxia.
    c. aprasia.
    d. anemia.

13. Which of the following is not assessed by the Brazelton Neonatal Behavior Assessment Scale?
    a. intermodal perception
    b. motor behaviors
    c. response to stress
    d. physiological control

14. Jan has a gestational age of 43 weeks. She would be considered to be
    a. premature.
    b. full-term.
    c. postmature.

15. Which of the following is the best statement regarding the outcome of premature infants?
    a. Most infants who are born prematurely will have major intellectual impairments later in life.
    b. About half of the infants born in the lowest birth weight categories will be neurologically impaired; the other half will develop normally.
    c. Many children who were born prematurely will never catch up to their full-term peers in terms of height and weight.
    d. Most preterm infants will catch up to their full-term peers in terms of intellectual capabilities in the first few years.

16. Robin was born at 41 weeks gestation and weighed only 4.5 pounds. She is likely to
    a. be more seriously impaired than other infants born at 41 weeks.
    b. be as developed as other infants born at 41 weeks.
    c. have the same problems as premature infants.
    d. experience deficits later in life.

17. The neurologic prognosis is good for small-for-gestational age infants if
    a. asphyxia is avoided.
    b. they are born before 37 weeks.
    c. their blood sugar is low.
    d. all of the above

18. At birth, the average full-term baby in the U.S. is about _____ inches long and weighs about _____ pounds.
    a. 22; 5.5
    b. 22; 7
    c. 20; 5.5
    d. 20; 7

19. When babies are born, they are covered with a cheese-like substance called
    a. extraneous fur.
    b. vertex cartosa.
    c. vernix caseosa.
    d. bortex catono.

20. Jodie developed physiological jaundice as a newborn. She was probably treated by
    a. placing her in the NICU and closely monitoring her heart rate and breathing.
    b. maintaining her body temperature at a constant rate.
    c. placing her under fluorescent lights.
    d. bathing her in a warm saline bath.

21. In a newborn, the sense that is the least developed is
    a. vision.
    b. hearing.
    c. smell.
    d. taste.

22. By what age is visual acuity similar to that of adults?
    a. at birth
    b. by 1 month
    c. by 6 months
    d. by 1 year

23. A baby who can distinguish her mother's voice from other voices could be as young as
    a. a newborn.
    b. 1 month old.
    c. 3 months old.
    d. 5 months old.

24. Paul's mother placed her finger in his palm when he was just a newborn, and Paul grasped it so tightly that he could be pulled up by it. This is called the
    a. Moro reflex.
    b. Babkin reflex.
    c. rooting reflex.
    d. Palmar grasp reflex.

25. Which of the following is a true statement about the neonate's brain?
    a. The neonate has the same amount of neurons present in the brain as do adults.
    b. The dendrites of neurons are well developed, but other parts of the neurons will continue to grow.
    c. Most of the neurons in the newborn are already myelinated.
    d. The brain of the newborn is fairly immature, with some parts not yet functional.

ANSWER KEY

APPLICATIONS I

1. show
2. labor
3. Dick-Read
4. cesarean section
5. anoxia
6. amniotic sac
7. Lamaze
8. General
9. perineum; an episiotomy
10. local

APPLICATIONS II

1. Apgar score
2. small-for-gestational-age
3. asphyxia; hypoglycemia
4. postpartal depression
5. neurons
6. Leboyer method
7. Brazelton Neonatal Behavior Assessment Scale
8. circumcision
9. premature
10. reflexes

MULTIPLE CHOICE

1. a (p. 98)	9. c (p. 104)	17. a (p. 112)
2. b (p. 99)	10. d (p. 105)	18. d (p. 113)
3. d (p. 101)	11. a (p. 106)	19. c (p. 113)
4. b (p. 100)	12. b (p. 106)	20. c (p. 113)
5. c (p. 100)	13. a (p. 107)	21. a (p. 114)
6. c (p. 102)	14. c (p. 109)	22. c (p. 114)
7. a (p. 102)	15. d (p. 110)	23. a (p. 114)
8. c (p. 102)	16. b (p. 111)	24. d (p. 115)
		25. d (p. 116)

Chapter 5

PERSPECTIVES ON CHILD DEVELOPMENT

CHAPTER OUTLINE

I. Child Development as a Subject of Study -

Child development is a specialized discipline devoted to the understanding of all aspects of human development from birth to adolescence.

II. Historical Perspectives

　　A. Children as Miniature Adults - During the Middle Ages, childhood was not considered a separate stage of life and children were expected to be little adults.

　　B. Children as Burdens - Before modern birth control was available, children who were not wanted were often considered to be a burden, and were often placed in orphanages where many of them died.

　　C. Utilitarian Value of Children - Until the 20th century, children were often abusively used as child laborers; England passed the first child labor laws in 1832.

III. Early Philosophies Regarding the Moral Nature of Children

　　A. Original Sin -
　　　　1. According to the Christian doctrine of original sin, children are born sinful and rebellious and are in desperate need of redemption. Only God could save them from eternal damnation, but in the meantime, the role of the parents was to break the rebellious spirit through strict discipline.
　　　　2. In contrast, Bushnell emphasized that God's love and grace are mediated through caring parents; this view is the forerunner of modern concepts of child development.

　　B. Tabula Rasa: John Locke - According to Locke, children are born like blank slates, neither good nor bad, and experience determines how they will turn out.

    C. Noble Savages: Jean-Jacques Rousseau
       1. Children are endowed with a sense of right and wrong.
       2. There are 4 stages of development; infancy, childhood, late childhood, and adolescence, which parents should be sensitive to.
       3. Rousseau emphasized maturation.

IV. Evolutionary Biology

    A. Origin of Species: Charles Darwin - The human species has evolved over millions of years through the process of natural selection and the survival of the fittest. Darwin emphasized that humans are kin to all living things, that individual differences are important, that behavior is adaptive, and that the scientific method is important.

    B. Recapitulation Theory: G. Stanley Hall - The development of the growing child parallels the evolution of the human species. Unlike Darwin's theory, this theory is outdated.

V. Baby Biographies - During the 19th and early 20th centuries, children were studied by keeping biographical records of their behavior, but these records were not objective.

VI. Normative Studies

    A. The Contents of Children's Minds: G. Stanley Hall - Hall began the movement of normative studies by having children answer questionnaires about their lives.

    B. Growth Patterns: Arnold Gesell
       1. Gesell wrote volumes about typical motor development, social behavior, and personality traits.
       2. He emphasized the role of biological maturation in determining development.

    C. Intelligence Testing: Lewis Terman - Terman published the first widely used intelligence test in the U.S., the Stanford-Binet Intelligence scale.

VII. Modern Contributions to Child Development

    A. Research Centers - Interest in research in child development blossomed after World War I and a number of centers opened across the United States.

    B. John Dewey emphasized that the school is a laboratory of life. Children should learn through experiences, and education should emphasize physical, mental, and moral development.

C. Medical and Mental Health Practitioners
1. Medical and child guidance practitioners have provided much information about child development.
2. A new field, called developmental pediatrics, has emerged, which integrates both medical and psychological knowledge.

D. Today - Although we have considerable medical and psychological knowledge, we are still lacking in areas such as adequate child care for working parents, and prevention of child abuse and infant mortality.

LEARNING OBJECTIVES/ STUDY QUESTIONS

After reading this chapter, you should be able to:

1. Describe what the discipline of child development involves.

2. Describe the following historical perspectives on children's development:
   a. Children are seen as miniature adults.

   b. Children are considered to be burdens.

   c. There is a utilitarian value of children.

3. Describe the following early philosophies of the moral nature of children:
   a. Original sin -

   b. Tabula rasa -

   c. Noble savages -

4. Describe Darwin's concepts of natural selection and survival of the fittest. What aspects of Darwin's theory are still maintained today?

5. Describe Hall's recapitulation theory.

6. Describe some of the early studies that looked at child development.

7. Discuss how the study of child development has changed and modernized over the years. What still needs to be improved upon?

KEY TERMS

In your own words, provide a definition for each of the following terms.

1. Child development_____

2. Original sin_____

3. Tabula rasa_____

4. Noble savages_____

5. Maturation_____

6. Natural selection_____

7. Survival of the fittest_____

8. Recapitulation theory_____

9. Baby biographies_____

10. Normative Studies_____

11. Stanford-Binet Intelligence Scale_____

12. Developmental Pediatrics_____

APPLICATIONS

For each of the following, fill in the blanks with a term listed above.

1. A person who believes that children are born morally neutral and will develop based on their experiences in the world might agree with Locke that children are a _____.

2. The new field which has emerged that combines the knowledge of developmental psychology with pediatric medicine is called _____.

3. During the late 19th century, researchers such as Darwin attempted to study child development by keeping _____.

4. Rousseau believed that children know what is right and wrong and thus he considered them to be _____.

5. The first intelligence test used in the United States was the
_____.

6. The idea that certain species survived because they had adaptive characteristics is the theory of _____.

7. Some researchers, such as Gesell, who wanted to look at typical development conducted _____.

8. A person who believes that children are born sinful and need to be saved may subscribe to the doctrine of _____.

9. The _____ theory suggests that children playing survival games parallels a period in history in which humans were cave-dwelling hunters and fishers.

10. _____ is the unfolding of the genetically determined patterns of growth and development.

SELF-TEST MULTIPLE CHOICE QUESTIONS

Circle the best answer for each question.

1. During the Middle Ages, childhood was <u>not</u> considered to be
   a. a time to spend learning to read and write.
   b. a period in which sexuality could be explored.
   c. a separate stage of life from adulthood.
   d. any of the above

2. Before birth control became readily available, children were
   a. considered to be a blessing.
   b. often murdered because infanticide was not considered a crime.
   c. considered to be a burden on the household.
   d. often put into orphanages where they thrived.

3. The first child labor laws were not passed until
   a. 1642.
   b. 1756.
   c. 1789.
   d. 1832.

4. Belief in the doctrine of original sin may have led parents to
   a. nurture their children in a warm and caring fashion.
   b. use harsh punishments for their children.
   c. reject Christian beliefs in God as the savior.
   d. believe that their children were inherently good.

5. Horace Bushnell believed that
   a. all children were born inherently sinful and were in need of redemption.
   b. children should have caring and loving parents who guide them in what is right and wrong.
   c. parents' major role in moral development is praying for their children's salvation.
   d. parents should use strict punishment to break a child of sinfulness.

6. Theorists or philosophers who believe that children are born as a tabula rasa believe that children are born
   a. inherently good.
   b. inherently bad.
   c. morally neutral.
   d. with good table manners.

7. The first person to emphasize the role of maturation in development was
   a. Rousseau.
   b. Darwin.
   c. Locke.
   d. Terman.

8. The survival of giraffes and the demise of animals with shorter necks who lived on plains where the major source of food was from the leaves on the tall trees is an example of
   a. recapitulation theory.
   b. maturational theory.
   c. tabula rasa.
   d. the process of natural selection.

9. According to which theory does the child reenact the animal stage of development?
   a. natural selection
   b. maturational theory
   c. recapitulation theory
   d. survival of the fittest theory

10. Which theorist believed that socialization could not overcome the effects of maturation on development?
    a. G. Stanley Hall
    b. Arnold Gesell
    c. Lewis Terman
    d. Charles Darwin

11. The first intelligence tests, designed by Binet in Paris, were intended to
    a. identify retarded children in the school system.
    b. determine which children should enter private schools for gifted children.
    c. identify which children needed special help in social adjustment to new situations.
    d. all of the above

12. Although considerable knowledge has been gained in the area of child development, there is still a need for resolving problems of
    a. inadequate child care.
    b. child sexual and physical abuse.
    c. infant mortality.
    d. all of the above, and more.

ANSWER KEY

APPLICATIONS

1. tabula rasa
2. developmental pediatrics
3. baby biographies
4. noble savages
5. Stanford-Binet Intelligence Scale
6. natural selection
7. normative studies
8. original sin
9. recapitulation
10. Maturation

MULTIPLE CHOICE

1. c (p. 126)
2. c (p. 126)
3. d (p. 129)
4. b (p. 129)
5. b (p. 129)
6. c (p. 130)
7. a (p. 130)
8. d (p. 130)
9. c (p. 131)
10. b (p. 133)
11. a (p. 133)
12. d (p. 137)

Chapter 6

PHYSICAL DEVELOPMENT

CHAPTER OUTLINE

I. Physical Growth

   A. Body Height and Weight
      1. From birth to 1 year, there is rapid but decelerating growth.
      2. From 1 year until the onset of puberty, growth is more steady and linear.
      3. During puberty, growth spurts are evident for boys and girls.

   B. Individual Differences - There are wide individual differences in growth patterns, influenced by heredity, nutrition, general health, and cultural and ethnic differences.

   C. Body Proportions - Not all parts of the body grow at the same rate.
      1. Cephalocaudal principle - Growth proceeds downward from the head to the feet.
      2. Proximodistal principle - Growth proceeds from the center of the body outward to the extremities.

   D. Organ Systems - The lymphoid system, the reproductive system, and the central nervous system do not follow the general pattern of growth.

   E. Brain Growth and Nerve Maturation
      1. There is an increase in myelinization of individual neurons.
      2. The cerebral cortex contains the higher brain centers controlling sensory, motor, and intellectual functions. Its two hemispheres each perform specialized functions.
      3. The motor area of the cortex matures first, followed by the sensory area, and then the association areas. Broca's area is involved in speech, and Wernicke's area is involved in understanding language.

   F. Teeth Eruption Times - The timing of teeth eruption is somewhat variable, depending on heredity and nutrition.

II. Motor Development

    A. Gross Motor Skills and Locomotion During Infancy - Motor development is dependent primarily on overall physical maturation, especially on skeletal and neuromuscular development.

    B. Fine Motor Skills During Infancy - These are the skills using the smaller muscles of the body, such as reaching and grasping.

    C. Gross Motor Skills of Preschool Children - Preschool children show increased skill and mastery of their bodies in performing physical feats.

    D. Fine Motor Skills of Preschool Children - Preschool children become more adept at performing activities that involve a high degree of small muscle and eye-hand coordination.

    E. Handedness - Handedness develops slowly and is not always consistent in the early years.

    F. Changes During the School Years -
       1. There is an increase in motor abilities as their bodies continue to grow, as well as improvement in fine motor skills.
       2. Girls are more physically mature than same-age boys. Prior to puberty, many differences in motor skills are due to differential expectations and experiences of boys and girls.

    G. Physical Fitness - Today's schoolchildren are less physically fit because they are not active enough.

III. Physically Handicapped Children

    A. Speech Handicapped Children - Speech handicaps may be due to congenital malformations, hearing, neurological, or developmental problems.

    B. Hearing Handicapped Children - Hearing problems may not be discovered until the child is one or two years old.

    C. Visually Handicapped Children - Blindness may be congenital or develop later. Visually handicapped infants usually lag behind sighted infants in locomotion.

    D. Children with Skeletal-Orthopedic-Motor Skills Handicaps - These types of handicaps range from almost complete disability to minor dysfunctions.

    E. Handicapped children may be subject to cruel teasing. By law, all handicapped children must receive public education in the least restrictive environment.

IV. Perceptual Development

   A. Depth Perception - Depth perception develops very early in infancy.

   B. Perception of Form and Motion - These abilities develop and change over the first few years of life.

   C. Perception of the Human Face - Infants prefer to look at human faces over objects.

   D. Auditory Perception - The abilities to localize sounds and to detect the gaps between words are important for auditory perception.

V. Nutrition

   A. Breast-Feeding vs. Bottle-Feeding - Breast-feeding declined rapidly in popularity over this century, but recently has been gaining in popularity again.
      1. Advantages - Breast milk is the most nutritious food available, contains antibodies that immunize the infant from disease, is convenient and may be pleasurable for the mother.
      2. Disadvantages - It can limit the mother's freedom, exclude the father from feeding, and some drugs and chemicals can be passed to the baby.

   B. Dietary Requirements - A balanced diet is necessary for good health and vigor.

   C. Obesity - Obesity can be due to many factors, such as heredity, eating habits, activity level and psychological factors.

   D. Malnutrition - An inadequate diet, such as in the extreme case of marasmus, may lead to mental retardation and other serious problems. Kwashiorkor results when there is a protein deficiency.

VI. Sleep

   A. Needs, Habits, and Disturbances
      1. If infants are comfortable, they will get the amount of sleep they need; in contrast, 2-year-olds won't. Resistance to going to bed peaks between 1 to 2 years, and may be due to separation anxiety.
      2. Young children may experience nightmares, night terrors, or sleepwalking.

   B. Sleeping with Parents - Some authorities say that it's better not to take children into bed with them, even if the child wakes up frightened at night.

VII. Health Care

    A. Health Supervision of the Well Child - Supervision by medical personnel should include instruction of parents in child development, routine immunizations, early detection of disease, and early treatment of disease.

    B. Education - At a young age, children should be taught about good health habits and hygiene, as well as proper nutrition and exercise.

VIII. Sexual Development

    A. Infancy
      1. The infant's capacity for sexual response begins early, but the infants are too young to be consciously aware of the arousal.
      2. Infants begin to discover their bodies during the first year of life.

    B. Early Childhood - Children are curious about their own bodies, as well as others' bodies.

    C. Middle Childhood - Sexual experimentation probably increases during these years, but it becomes more covert because it is less accepted by society.

IX. Sexual Abuse of Children - Sexual abuse is most likely to occur in the child's home or the home of the molester. The abuser is most often known to the child. Sexual abuse often isn't reported because the child feels afraid or guilty.

LEARNING OBJECTIVES/ STUDY QUESTIONS

After reading this chapter, you should be able to:

1. Describe how growth proceeds from birth to adolescence, and what factors may create individual differences.

2. Explain the principles of cephalocaudal and proximodistal growth.
   a. cephalocaudal -

   b. proximodistal -

3. Describe some of the changes that take place in the brain over development.

4. Discuss how gross motor skills and fine motor skills change over development.

5. Discuss the development of handedness.

6. Discuss the different types of physical handicaps and how handicapped children have to adjust to their situation.

7. Describe the development of depth perception, perception of form and motion, and face perception.

8. Discuss some of the factors that affect auditory perception.

9. Discuss the advantages and disadvantages of breast-feeding.

10. Discuss some of the factors responsible for obesity in children.

11. Describe the sleep habits and some of the sleep disturbances of young children.

12. Describe how children's sexual curiosity and experimentation change over age.

13. Discuss the characteristics of sexual abuse.

KEY TERMS

In your own words, provide a definition for each of the following terms:

1. Cephalocaudal principle_____
_____

2. Proximodistal principle _____

3. Myelinization _____

4. Cerebral cortex _____

5. Lateralization _____

6. Broca's area _____

7. Wernicke's area _____

8. Deciduous teeth and permanent teeth _____

9. Gross motor skills _____

10. Fine motor skills _____

11. Cross-modal perception _____

12. Gap threshold _____

13. Inanition _____

14. Marasmus _____

15. Kwashiorkor _____

16. Separation anxiety _____

17. Nightmares _____

18. Night terrors _____

19. Sleepwalking _____

APPLICATIONS

For each of the following, fill in the blank with one of the terms listed above.

1. A one-year-old who cries a lot when her parents put her down to sleep may be experiencing _____.

2. A child who wakes up screaming in the middle of the night is experiencing _____.

3. The _____ is divided into two hemispheres and controls higher order functions.

4. According to the _____ principle, the arms would develop before the fingers.

5. According to the _____ principle, the head would develop more rapidly than the feet.

6. A young child who has not received an adequate intake of all nutrients may suffer from _____.

7. Running and jumping are examples of _____.

8. Using a pencil is an example of a _____.

9. The area of the cortex that is involved with speech is _____; the area that is involved with understanding others' speech is _____.

10. The child's first teeth, which fall out during early childhood, are called the _____ teeth.

11. The specialization of the left hemisphere for language for most people is referred to as _____.

12. A child who has not received foods with adequate protein and who has a protruding belly and flaky skin is suffering from _____.

SELF-TEST MULTIPLE CHOICE QUESTIONS

Circle the best answer for each question.

1. A period of rapid but decelerating growth occurs
    a. during the first year of life.
    b. during the second year of life.
    c. from the first year through puberty.
    d. during the period of time right before puberty.

2. That infants are able to run and jump before they can grasp and effectively manipulate objects demonstrates
   a. the cephalocaudal principle.
   b. the proximodistal principle.
   c. the large-to-small principle.
   d. the gross-muscle to fine-muscle principle.

3. For which of the following organ systems does most growth occur in the early years of life?
   a. lymphoid system
   b. reproductive system
   c. central nervous system
   d. all of the above

4. Myelinization of neurons
   a. helps to speed up the transmission of nerve impulses.
   b. makes transmission of impulses more efficient.
   c. allows children to be more capable of complex motor activities.
   d. all of the above

5. The left cerebral hemisphere
   a. controls the left side of the body.
   b. is specialized for language.
   c. is specialized for music.
   d. is superior in recognizing patterns.

6. Lateralization
   a. may be apparent in a newborn.
   b. becomes weaker over development.
   c. is not complete until well into adulthood.
   d. is evidenced only by right-handed people.

7. The first area of the cortex to mature is the
   a. association area.
   b. sensory area.
   c. motor area.
   d. language area.

8. A person who had an injury to Wernicke's area would most likely
   a. not be able to speak.
   b. not be able to understand what other people were saying.
   c. not be able to recognize other people.
   d. not be able to remember what had happened during the accident.

9. Permanent teeth
    a. tend to appear in the back of the mouth first.
    b. tend to be smaller than deciduous teeth.
    c. tend to appear earlier in boys than girls.
    d. tend to appear earlier in girls than boys.

10. By the time a child is 6 months old, she would most likely be able to
    a. pull herself up to a standing position.
    b. stand alone.
    c. sit alone.
    d. creep.

11. A child who was just starting to reach for objects placed in front of him would probably be around
    a. 3 months old.
    b. 5 months old.
    c. 7 months old.
    d. 9 months old.

12. Which of the following is an example of a fine motor skill achieved by preschool-aged children?
    a. being able to alternate feet
    b. walking on a balance beam
    c. holding a glass with one hand
    d. drawing in proportion

13. Young girls, as compared to boys of the same age, are _____ physically mature.
    a. equally
    b. more
    c. less

14. Sex differences found for physical skills before puberty are most likely due to
    a. boys being bigger and stronger than girls.
    b. boys having quicker reaction times.
    c. girls being more mature.
    d. differential expectations of boys and girls.

15. Which of the following is the best statement concerning the physical fitness of children today as compared to children in the 1960's?
    a. Children today are less active.
    b. Children today have greater muscle strength.
    c. Children today have greater physical endurance.
    d. Children today have less body fat.

16. A child who is 18 months old is asked by her pediatrician to point to various simple objects and body parts in response to tape recorded names of objects. The doctor controls the volume that each word is spoken. This doctor is probably testing
    a. the speech reception threshold.
    b. the gap threshold.
    c. sound localization.
    d. expressive language.

17. A child with Duchenne muscular dystrophy
    a. will show symptoms at birth.
    b. will probably die by age 5.
    c. is most likely an infant girl.
    d. none of the above

18. Little Johnny was able to recognize a stuffed lion toy by feeling it, even though he had previously only seen it and not had the opportunity to touch it. This is an example of
    a. binocular vision.
    b. cross-modal perception.
    c. visual localization.
    d. unilateral perception.

19. We are able to figure out where sounds come from because
    a. our ears are on opposite sides of our heads.
    b. sounds arrive at one ear more quickly than at the other.
    c. sounds are louder in one ear than in the other.
    d. all of the above

20. One of the advantages that breast-feeding has over bottle-feeding is that
    a. it allows the father to participate more in the feedings.
    b. it is less painful for the mother.
    c. it is assured that no unwanted chemicals will be passed to the baby.
    d. antibodies that immunize the baby from disease are passed from the mother to the baby.

21. If a child's parents are obese, the child is more likely to be obese because
    a. parents teach their children bad eating habits.
    b. it is more likely the child will be born with more and larger fat cells.
    c. the child may have inherited the tendency to have a voracious appetite.
    d. all or any of the above

22. A child is most likely to have negative effects from eating sugar if
    a. her diet generally has adequate amounts of proteins.
    b. she has already eaten a lot of carbohydrates that day.
    c. her blood sugar is low.
    d. she is participating in a fun activity.

23. Which is the best statement regarding getting enough sleep?
    a. Young infants only need about 10 hours of sleep per day, and this amount increases over the first year.
    b. Two-year-olds are more likely to get the amount of sleep they need than are infants.
    c. Infants are more likely to get the amount of sleep they need than are two-year-olds.
    d. Infants are more likely than older children to follow a sleeping schedule.

24. According to the text, when a young child gets out of bed after being put to sleep, the best thing for a parent to do to break this child of that habit is
    a. let him sleep in his parents' bed for a little while.
    b. lie down in bed with him until he falls asleep.
    c. put him back in bed and wait in the hallway until he falls asleep.
    d. let him stay up a little bit longer because he may not have been tired enough to fall asleep.

25. For a three-year-old who has trouble distinguishing fantasy from reality, watching a scary movie may lead to
    a. nightmares.
    b. sleepwalking.
    c. increased nap-taking.
    d. all of the above

26. Immunizations
    a. can all be taken care of in the first year of life.
    b. cannot be given until the child is at least a year old.
    c. should be given throughout the lifespan depending on the type of immunization.
    d. are not that important to receive in the U.S. because the risk of disease is minimal.

27. When is the earliest time that a male can have an erection?
    a. in utero
    b. during the first 24 hours of life
    c. not until around two years of age
    d. not until after puberty

28. Children become curious about other children's bodies and boy-girl differences
    a. as infants.
    b. in early childhood.
    c. in middle childhood.
    d. as adolescents.

29. During middle childhood, sexual experimentation may _____ and it also becomes more _____.
   a. decrease; covert
   b. decrease; overt
   c. increase; covert
   d. increase; overt

30. An adult who sexually abuses a child
   a. is most likely a stranger to the child.
   b. is most likely known to the child.
   c. most likely will do it in a place with which the child is not familiar.
   d. most likely will do it in an outdoor setting.

ANSWER KEY

APPLICATIONS

1. separation anxiety
2. night terrors
3. cerebral cortex
4. proximodistal
5. cephalocaudal
6. marasmus
7. gross motor skills
8. fine motor skill
9. Broca's area; Wernicke's area
10. deciduous
11. lateralization
12. kwashiorkor

MULTIPLE CHOICE

1. a (p. 144)
2. b (p. 146)
3. c (p. 146)
4. d (p. 147)
5. b (p. 147)
6. a (p. 148)
7. c (p. 148)
8. b (p. 149)
9. d (p. 151)
10. c (p. 152)
11. b (p. 153)
12. c (p. 153)
13. b (p. 156)
14. d (p. 156)
15. a (p. 157)
16. a (p. 158)
17. d (p. 159)
18. b (p. 161)
19. d (p. 162)
20. d (p. 163)
21. d (p. 165)
22. b (p. 167)
23. c (p. 169)
24. c (p. 169)
25. a (p. 169)
26. c (p. 172)
27. a (p. 170)
28. b (p. 173)
29. c (p. 173)
30. b (p. 173)

Chapter 7

COGNITIVE DEVELOPMENT

CHAPTER OUTLINE

I. Language

    A. Language and Communication - Human infants can communicate through reflexive actions and nonverbal behavior such as facial expressions. Culture affects maternal speech to children and reflects child-rearing goals.

    B. Elements and Rules of Language - The basic elements of language are phonemes, morphemes, syntax, semantics, and pragmatics.

    C. Theories of Language Development
       1. Biological theory - According to the nativist view, children inherit a predisposition to learn language at a certain age.
       2. Learning theory - Language is learned like other behavior is learned, through imitation, conditioning, and reinforcement.
       3. Cognitive theory - According to Piaget, language develops out of mental images. In contrast, the Whorfian hypothesis contends that language influences thought.
       4. Interactionist theory - Maturation and experience are both emphasized.

    D. Influences on Language Development - Both biological maturation and environmental influences affect language development. For example, language abilities of culturally-deprived children can be improved through reading.

    E. Sequence of Language Development
       1. Prelinguistic period - All children follow the same timetable and sequence from crying, to cooing, to babbling.
       2. First spoken words - At about 10 months, infants use holophrases, and by 18 months, may know between 3 to 50 words, mostly referring to objects.
       3. Two-word utterances - Duos usually begin between 18-24 months.
       4. Telegraphic speech - utterances which exclude unnecessary words.
       5. Sentences - By 2 1/2 to 4 years, children use simple sentences, and by 6-7

years, children's speech resembles that of adults.

F. Vocabulary and Semantics - Vocabulary grows from about 50 words at age 2 to between 8,000 and 14,000 at age 6. By 18 months, children are able to categorize objects.

G. Grammar - Children show some knowledge of grammar by the time they begin using sentences.

H. Pragmatics - develop during the elementary school years.

I. Gender and Communication Patterns - Boys' use of language emphasizes dominance, whereas girls emphasize cooperation.

J. Talking About Feelings - One research study suggested that mothers tend to use comforting language, whereas siblings tend to focus on their own feelings.

K. Bilingualism
   1. If the child's first language is a minority language, learning the majority language will detract from the first, and it is better to be taught in their first language.
   2. If the first language is the majority language, then learning a second language is largely an additive experience.

L. Stuttering - Stuttering appears to have a genetic base.

II. Approaches to the Study of Cognition - There are 3 basic approaches: Piagetian, information-processing and psychometric.

III. A Piagetian Perspective - Piaget described 4 stages of cognitive development:

A. Sensorimotor Stage (birth - 2 years) - involves learning to respond through motor activity to the stimuli that are presented to the senses; there are six substages.
   1. The concept of object permanence develops.
   2. Even very young infants can imitate or copy simple behaviors of others.
   3. What and how much children learn is dependent upon environmental stimulation.

B. Preoperational stage (2 to 7 years) - Children do not think logically, but they can think symbolically or representationally.
   1. During this time, children are able to engage in symbolic play.
   2. Their thought processes are not always like adults in that they use transductive reasoning, syncretism, egocentric thinking, animism, and centration.

3. They lack the abilities to conserve and to classify and have problems with irreversibility.
4. During this stage, children come to have better understanding of space, proportions, causality, beliefs, and kinship.

C. Concrete Operational Stage (7 to 11 years) - Children are better at logical reasoning, but only at a concrete level.
1. They have difficulty with contrary-to-fact reasoning and fail to test hypotheses.
2. They can arrange objects into hierarchical classifications and comprehend class inclusion relationships.
3. They are successful at seriation and conservation tasks.
4. They can perform combinativity, reversibility, associativity and identity or nullifiability.

D. Vgotsky's Theory of Cognitive and Language Development - In this view, mental functioning is derived primarily from social and cultural influences. It is important to understand the child's zone of proximal development.

IV. Information Processing - describes the way children obtain information, remember it, retrieve it, and use it in solving problems.

A. Stimuli - Research has shown the importance of stimulation in learning.

B. Habituation - When infants get used to a sound or sight, it loses its novelty and the infants lose interest in it. This process has been used to measure perception, memory, neurological health, and intelligence.

C. Selective Attention - There are dramatic increases in selectivity with age.

D. Memory
1. Infant memory - Newborns have some memory ability, but it is very short-lived, and early memories are not permanent.
2. Memory capacity and storage - The process of remembering involves sensory storage, short-term storage, and long-term storage.
3. Metamemory - consists of knowledge of memory strategies that people employ to learn and remember information, which increase from preschool through adolescence.
4. Organization can facilitate children's memory of where things are.

V. Intelligence - the psychometric approach

A. Views of Intelligence
1. Binet - Intelligence is a general capacity, described in terms of mental age.
2. Stern originated the term intelligence quotient (IQ).

3. Spearman - Besides a general intellectual factor ("g"), there are also specific abilities.
4. Others, such as Gardner, have also identified multiple dimensions or factors of intelligence.
5. Sternberg's triarchic theory of intelligence includes componential, experiential, and contextual intelligence.
6. Cattell described crystallized and fluid dimensions of intelligence.

B. Intelligence Tests - Some of the commonly used tests are the Stanford-Binet and the Wechsler Scales.

C. Critique of IQ and IQ tests -
1. IQ tests do a pretty good job of predicting school performance, and a better job of predicting job success in some occupations.
2. At age 2, tests cannot predict later scores, but by age 5, future scores are more predictable, though there are individual differences in stability.
3. Test results can be influenced by test anxiety, interest in the tasks, and rapport with the test-giver.
4. The major criticism of IQ tests is that they are culturally biased in favor of white middle-class children.

D. IQ and Race - Differences in IQ scores between races may be due to social class differences and cultural biases of the tests.

E. Infant Intelligence and Measurement - Some of the measures used for infants include the Developmental Quotient and Bayley's Scales of Infant Development, but parental IQ and educational level and measures of habituation are better predictors of later IQ scores. The Fagan Test of Infant Intelligence uses visual recognition memory as a screening device.

F. Early Intervention - High quality programs for disadvantaged children can have lasting and valuable effects, but sometimes can create other types of problems.

G. Mental Retardation - Mentally retarded individuals may be classified as borderline, mildly retarded, moderately retarded, severely retarded or profoundly retarded.

VI. School

A. Early Childhood Education - Some of the different types of programs include: Nursery schools, Montessori schools, Group-care home, and Day-care centers. The key to success of these programs is the quality of the services provided.

B. American Education - The American education system is frequently criticized for being mediocre. For example, Japanese and Taiwanese students do much better on mathematics exams than do students in the U.S.

C. Successful Schools - Successful schools in the U.S. emphasize academic excellence, pay attention to the needs of individual students, emphasize no-nonsense discipline, and employ great teachers.

D. Achievement
1. Heredity can affect achievement because it is an important factor in intelligence.
2. Children with learning disabilities, such as dyslexia, have problems in specific areas such as reading or arithmetic.
3. Achievement motivation, dysfunctional family relationships, and socio-cultural influences can all affect achievement.

LEARNING OBJECTIVES/ STUDY QUESTIONS

After reading this chapter, you should be able to:

1. Describe the five basic elements of language:
   a.

   b.

   c.

   d.

   e.

2. Discuss the following theories of language development:
   a. biological theory -

   b. learning theory -

   c. cognitive theory -

   d. interactionist theory -

3. Briefly discuss both the biological and environmental influences on language development.

4. Briefly summarize the sequence of language development from early infancy through early childhood.

5. Briefly discuss the development of vocabulary, grammar, and pragmatics.

6. Describe gender differences in communication patterns.

7. Discuss the effects of learning a second language on language development.

8. Describe Piaget's first three stages of cognitive development, including what children can and cannot do during these stages:
   a.

b.

　　c.

9. Discuss Vgotsky's view of cognitive and language development.

10. Describe some of the factors which are important in the information-processing approach.

11. Discuss the three stages of memory.
　　a.

　　b.

　　c.

12. Define metamemory and describe some mnemonic strategies.

13. Discuss some of the different definitions or views of intelligence.

14. Discuss some of the measures used for measuring IQ for infants and for children. Provide a critique of how effective and fair these measures are.

15. Describe the five categories of mental retardation.

16. Describe some of the different types of early childhood education programs.

17. Discuss the status of the American education system, including a description of what makes some schools better than others.

18. Describe some of the factors that influence achievement.

KEY TERMS I

In your own words, provide a definition for each of the following terms:

1. Phoneme _____

2. Morpheme _____

3. Syntax _____

4. Semantics _____

5. Pragmatics _____

6. Language acquisition device _____

7. Cooing _____

8. Babbling _____

9. Holophrases _____

10. Motherese _____

11. Duos _____

12. Telegraphic speech _____

13. Grammar _____

14. Object permanence _____

15. Deferred imitation _____

APPLICATIONS I

For each of the following, fill in the blank with a term listed above.

1. A child who says, "Mommy went bye-bye" rather than "Went bye-bye Mommy" is showing some knowledge of _____ or _____.

2. Two-word utterances are referred to as _____.

3. Knowing that the word "cat" refers to four-legged animals that purr and meow is an

issue of _____.

4. A three-month-old who squeals and gurgles is _____.

5. "Kitty drink milk" is an example of _____.

6. An infant who watches her big brother stick his tongue out, and then later on, when he is no longer around, sticks her own tongue out may be engaging in _____.

7. The sound "ch" is one of the 44 _____ in the English language.

8. A child who knows to speak respectfully to her teacher is aware of the _____ of language usage.

9. A seven-month-old who repeats the syllable "da" over and over is _____.

10. The first spoken words that a child uses are _____.

KEY TERMS II

In your own words, provide a definition for each of the following terms:

1. Symbolic play_____

2. Transductive reasoning_____

3. Inductive reasoning_____

4. Deductive reasoning_____

5. Syncretism_____

6. Egocentrism_____

7. Animism_____

8. Centration_____

9. Conservation_____

10. Classification_____

11. Irreversibility_____

12. Hierarchical classification _____

13. Class inclusion relationships_____

14. Serialization_____

15. Combinativity_____

16. Reversibility_____

17. Associativity_____

18. Identity or nullifiability_____

19. Zone of proximal development_____

20. External speech_____

21. Internal speech_____

22. Egocentric speech_____

APPLICATIONS II

For each of the following, fill in the blank with one of the terms listed above.

1. The tendency to focus on only one aspect of a situation and to ignore the other aspects is called _____.

2. During the last storm, the lights in Marta's house went off. During the next storm, Marta expects the lights to go off again. Marta is engaging in
_____.

3. A child who understands that all the boys in her class and all the girls in her class equals all of the children in her class understands the operation of
_____.

4. A child is shown two balls of clay of the same size. Then one ball is flattened out and the child is asked which of them is bigger. This is an example of a
_____ task.

5. A child who pretends that she and her dog are on a ship that is sailing far away is engaging in _____.

6. A child who understands that dogs are a type of animal understands _____ and _____.

7. A child who understands that if she is given two apples and then two apples are taken away, she won't have any apples understands the operation of _____ or _____.

8. A 4-year-old who thinks to herself about what she will do later that afternoon can be said to have _____ speech.

9. A child who says, "That flower wants me to pick it." is engaging in _____.

10. A child who knows that dogs bark, and then sees a new dog in her neighborhood and reasons that this dog will also bark is engaging in _____ reasoning.

11. A child who takes all the almonds out of a bowl of mixed nuts and puts them in one pile, and then takes the walnuts and puts them into another pile, can be said to be successful at _____.

KEY TERMS III

In your own words, provide a definition for each of the following terms:

1. Habituation _____

2. Infantile amnesia _____

3. Sensory storage _____

4. Short-term storage _____

5. Long-term storage _____

6. Recall _____

7. Recognition _____

8. Metamemory _____

9. Mnemonic strategies _____

10. Chunking _____

11. Method of loci _____

12. Mental age (MA) _____

13. Chronological age (CA) _____

14. Intelligence quotient _____

15. Crystallized intelligence _____

16. Fluid intelligence _____

17. Developmental quotient (DQ) _____

18. Mental retardation _____

19. Learning disabilities _____

20. Dyslexia _____

APPLICATIONS III

For each of the following, fill in the blank with one of the terms listed above.

1. Mental age is the level of development in relation to _____.

2. _____ storage has a limit of about 7 items which can be stored.

3. A child who has played with her green dumptruck so many times that she is bored with it can be said to have _____ to it.

4. Mary has an IQ of 110, but she has a lot of trouble learning even simple mathematics. Mary may have a _____.

5. People often use _____ strategies, such as repeating the information, to remember things.

6. A person's _____ is calculated by dividing the mental age by the chronological age and multiplying by 100.

7. The inability of adults to remember events in their lives that took place before they were three years old is called _____.

8. According to Cattell, _____ intelligence is a person's ability to think and reason abstractly.

9. Knowing how it was that you remembered something is part of _____.

10. A child with an average IQ who reads letters from right to left and reads the letter "b" as a "d" may have _____.

11. A child who remembers a group of objects by remembering all the animals together and then all the vegetables together is using the strategy of _____.

12. A child who remembers what furniture is in her house by mentally walking through the different rooms is using the strategy of _____ to aid in retrieval.

SELF-TEST MULTIPLE CHOICE QUESTIONS

Circle the best answer for each question.

1. The aspect of language that deals with the meaning of words and phrases is
    a. morphemes.
    b. syntax.
    c. semantics.
    d. pragmatics.

2. Which theory maintains that infants are born with a language acquisition device?
    a. biological theory
    b. learning theory
    c. cognitive theory
    d. interactionist theory

3. Which of the following has been suggested as a possible biological influence on language development?
    a. amount of time mothers spend interacting with their children
    b. the child's temperament
    c. the content of what parents communicate to their children
    d. the social and economic status of the family

4. Which of the following is a valid argument for why babbling is not learned through imitation?
   a. Children all over the world begin to babble at around the same age.
   b. All children follow through the same sequence of preverbal communication.
   c. Even deaf children begin to babble.
   d. all of the above

5. Children who are three years old should be speaking
   a. in holophrases.
   b. in duos.
   c. in three to five word sentences.
   d. in six to eight word sentences.

6. Which of the following words is most likely to be learned first?
   a. dog
   b. bark
   c. animal
   d. beagle

7. Pragmatics is an aspect of language that develops during
   a. infancy.
   b. the preschool years.
   c. the elementary school years.
   d. adolescence.

8. Which of the following is the best statement about gender differences in communication patterns?
   a. Girls are more likely than boys to be demanding.
   b. Girls are more likely than boys to be direct in their language.
   c. Girls are more likely than boys to emphasize mutual cooperation.
   d. Girls are more likely than boys to use domineering language.

9. A child lying in her crib accidentally hits the mobile that is hanging over her. She smiles gleefully and does it over and over again. According to Piaget, this is called a
   a. primary circular reaction.
   b. secondary circular reaction.
   c. tertiary circular reaction.
   d. external circular reaction.

10. A child who knows that a ball continues to exist even when it rolls under the couch out of his sight can be said to have of a concept of
    a. circular reactions.
    b. secondary schemes.
    c. object independence.
    d. object permanence.

11. Children develop the ability to imagine doing something without having to actually do it during which of Piaget's stages?
    a. sensorimotor
    b. preoperational
    c. concrete operational
    d. formal operational

12. A child who doesn't want to say that a rock is ugly because he says that he doesn't want to hurt the rock's feelings is engaging in
    a. animistic thinking.
    b. egocentric thinking.
    c. centration.
    d. syncretism.

13. A preoperational child cannot
    a. engage in symbolic play.
    b. use words to represent objects.
    c. understand that some operations are reversible.
    d. all of the above

14. A child who is able to solve conservation problems but cannot understand hypothetical or contrary-to-fact propositions is in which of Piaget's stages?
    a. sensorimotor
    b. preoperational
    c. concrete operational
    d. formal operational

15. John has a workbook in which each successive problem is harder to solve. When working by himself, John can solve up to problem #15, but when his teacher sits down with him and gives him some guidance, he can solve up to #26. This task could be used to measure
    a. the zone of proximal development.
    b. collective memory.
    c. inner speech.
    d. distal problem-solving abilities.

16. Memory is first apparent
    a. in the early weeks of life.
    b. at around 3 months.
    c. at around 6 months.
    d. after the first year.

17. Information can be remembered for up to about 30 seconds in
    a. sensory storage.
    b. short-term storage.
    c. long-term storage.
    d. retroactive storage.

18. The short-term memory capacity is fairly well-developed by
    a. early childhood.
    b. adolescence.
    c. early adulthood.
    d. late adulthood.

19. A child is given a grocery list. He tries to remember it by grouping all of the vegetables together, then all of the dairy products together and all of the snacks together and then repeating them all out loud several times. He is using
    a. mnemonic strategies.
    b. the strategy of rehearsal.
    c. the strategy of organization.
    d. all of the above

20. If a 4 year old child has a mental age of 6, what is her intelligence quotient?
    a. 100
    b. 120
    c. 150
    d. 170

21. According to Sternberg, the aspect of intelligence that includes the ability to acquire and store information is
    a. componential intelligence.
    b. experiential intelligence.
    c. logical intelligence.
    d. contextual intelligence.

22. The ability to think and reason abstractly is referred to by Cattell as
    a. fluid intelligence.
    b. crystallized intelligence.
    c. formal intelligence.
    d. intrapersonal intelligence.

23. A high score on an IQ test is predictive of
    a. success at a job that involves getting along with others.
    b. success at a job that involves taking risks.
    c. success at a job that involves academic skills.
    d. all of the above

24. The System of Multicultural Pluralistic Assessment (SOMPA) was designed to
    a. assess both general knowledge and reasoning abilities.
    b. eliminate the cultural bias that exists in most intelligence tests.
    c. see if people of different races and ethnic backgrounds differed in terms of IQ.
    d. assess current abilities rather than potential.

25. Which of the following is the best predictor in infancy of a child's later IQ?
    a. the child's DQ
    b. scores on Bayley's Scales of Infant Development
    c. Gesell's measures of motor, language, adaptive, and social abilities
    d. parental IQ and educational level

26. A child with an IQ of 55 would be classified as having
    a. borderline mental retardation.
    b. mild mental retardation.
    c. moderate mental retardation.
    d. severe mental retardation.

27. The most important factor for evaluating early childhood education programs is
    a. the type of program (for example, Montessori vs. nursery schools).
    b. the size of the center.
    c. the quality of the program.
    d. the number of children in the program.

28. According to a study by Stevenson and others, American children, in comparison to Japanese and Taiwanese children,
    a. score similarly on math and science exams.
    b. spend more time in school.
    c. receive more instruction from their teachers.
    d. spend more time in irrelevant activities.

ANSWER KEY

APPLICATIONS I

1. grammar; syntax
2. duos
3. semantics
4. cooing
5. telegraphic speech
6. deferred imitation
7. phonemes
8. pragmatics
9. babbling
10. holophrases

APPLICATIONS II

1. centration
2. syncretism
3. combinativity
4. conservation
5. symbolic play
6. hierarchical classification; class inclusion relationships
7. identity; nullifiability
8. inner
9. animism
10. deductive
11. classification

APPLICATIONS III

1. chronological age
2. Short-term
3. habituated
4. learning disability
5. mnemonic
6. intelligence quotient
7. infantile amnesia
8. fluid
9. metamemory
10. dyslexia
11. chunking
12. the method of loci

MULTIPLE CHOICE

1. c (p. 184)
2. a (p. 184)
3. b (p. 186)
4. d (p. 187)
5. c (p. 190)
6. a (p. 190)
7. c (p. 192)
8. c (p. 193)
9. b (p. 196)
10. d (p. 196)
11. b (p. 197)
12. a (p. 198)
13. c (p. 199)
14. c (p. 201)
15. a (p. 202)
16. a (p. 204)
17. b (p. 205)
18. b (p. 206)
19. d (p. 207)
20. c (p. 208)
21. a (p. 209)
22. a (p. 210)
23. c (p. 211)
24. b (p. 213)
25. d (p. 214)
26. b (p. 216)
27. c (p. 218)
28. d (p. 221)

Chapter 8

EMOTIONAL DEVELOPMENT

CHAPTER OUTLINE

I. Attachment

    A. Meaning and Importance
       1. Attachment is the emotional link between an adult and a child which includes the desire to maintain contact.
       2. The formation of attachments is important to a child's total development.
       3. There is an intergenerational influence on the formation of attachment relationships.

    B. Multiple Attachments - Children can form attachments to more than one person, but this doesn't mean that caregivers can be constantly changed.

    C. Specific Attachments - Attachments to specific persons do not develop until about 6 or 7 months of age.
       1. Maturation - Before attachment can occur, infants must learn to distinguish humans from inanimate objects, to distinguish between different people, and to develop a specific attachment to one person.
       2. Specific attachments are at their maximum from 12 to 18 months, and then attachment behaviors decrease over the course of the second year.

    D. Nonattached Children and Insecure Attachment
       1. Nonattached children may make no distinction between their own parents and other adults.
       2. Insecurely attached children, particularly girls, can be excessively dependent on their parents. Insecure boys may show aggressive and attention-seeking behaviors.

    E. Separation Anxiety
       1. Symptoms - Signs of separation anxiety, such as crying when the parent leaves, vary according to individual differences, age, and the length of separation.

2. With long-term separation, children initially protest, then reach a period of despair, followed by detachment and withdrawal.
3. Age factors - Distress over separation is greatest after 6 months, until about 3 years of age, but even preschool and school age children can be affected.

F. Reunion Behavior - Some children become very dependent and possessive; others may become angry.

G. Strangers - Fear of strangers usually begins at about 6 or 7 months of age and increases until about 2 years, after which it declines.

H. Baby-Sitters and Substitute Caregivers - The question of the long-term effects of substitute care on children is a very controversial one. The quality of the care is an important factor; high-quality daycare can have positive effects.

II. Development of Trust and Security

A. Theoretical Perspectives
1. Erikson suggested that the core of personality is formed in infancy as infants learn basic trust.
2. Margaret Mahler emphasized the importance of the mother-child relationship.

B. Requirements for the Development of Trust and Security in Infants - Children should receive regular and adequate feedings, get sufficient sucking, experience cuddling, physical contact and love from their parents.

C. Some Causes of Distrust and Insecurity
1. Parental Deprivation - The longer the deprivation, the more pronounced the effects.
2. Tension - Being cared for by anxious parents can lead to emotional insecurity.
3. Exposure to frightening experiences may lead to insecurity if it is traumatic.
4. Frequent disapproval and criticism may make children unsure of themselves.
5. Children who are overprotected may become anxious.
6. Children who are overindulged may not be prepared for dealing with frustrations later in life.

D. Child Abuse - Abused children are physically and emotionally harmed and often have trouble with social relationships.

E. Children of Alcoholics - These children are at high risk for behavioral and emotional problems.

III. Development of Emotions

    A. Components - Emotions involve the basic components of: stimuli, feelings, physiological arousal, and behavioral response.

    B. Functions - Emotions are adaptive, are a means of communication, and are important in social relationships and sociomoral development. They are powerful motivators, as well as a source of pleasure or of pain.

    C. Basic Emotions - The basic emotions which can be recognized by people all over the world are: happiness, sadness, anger, surprise, disgust and fear. It is more difficult to identify these emotions in infants.

    D. Children's Fears - Children are born with the ability to fear loud noises and falling. They develop other fears, some through actual experiences or because of limited understanding.

    E. Timetable of Development - Izard claims that emotions develop according to a biological timetable, but others suggest that all emotions can be experienced at birth.

    F. Environmental and Biological Influences in Development - Biology defines the broad outline and limits of emotional development, while environmental influences modify that evolution.

IV. Differences in Temperament

    A. Personality and Temperament
      1. Personality is the sum total of the physical, mental, emotional, and social characteristics of an individual; it is a global concept that is not static.
      2. Temperament refers to relatively consistent, basic dispositions inherent in people which influence their behavior. The expression of temperament is influenced by the environment.

    B. Components and Patterns of Temperament
      1. Buss and Plomin specify three traits: emotionality, activity, and sociability.
      2. Thomas and Chess specify nine components: rhythmicity, activity level, approach/withdrawal, adaptability, sensory threshold, mood, intensity, distractibility, and persistence. They found three temperament patterns: easy, difficult, and slow-to-warm up.

V. Development of Self, Autonomy, Self-Concept, and Self-Esteem

    A. Self-Awareness - Children begin to understand their separateness from others.

B. Autonomy - As the self emerges, children begin to want some independence.

C. Separation-Individuation - According to Mahler, at about 5 months, until about 3 years, a period of separation-individuation occurs during which the infant gradually develops a self apart from the mother.

D. Self-Definition and Self-Concept - By 3 years, children describe themselves in exaggerated and positive terms, but by the middle elementary school years, their descriptions become more realistic.

E. Self-Reference and Self-Efficacy - the perception of how effective the self is.
   1. Harter developed a scale that measures general self-worth, social skills, cognitive skills, and physical skills.
   2. Bandura suggested that children's judgement of self-efficacy stems from personal accomplishments, their comparisons to others, persuasion, and the person's arousal level.

F. Self-Esteem - how children feel about themselves; The primary sources of self-esteem are emotional relationships with parents, their social competence with peers, their intellectual abilities and the attitudes of society toward them.

LEARNING OBJECTIVES/ STUDY QUESTIONS

After reading this chapter, you should be able to:

1. Define "attachment" and describe to whom infants can become attached.

2. Describe nonattached and insecurely attached children.

3. Discuss separation anxiety and the long-term effects of separation on children.

4. Describe stranger fear and children's reactions to substitute caregivers.

5. Discuss some of the requirements for the development of trust and security in infants.

6. Discuss some of the causes of distrust and insecurity. What are the effects of child abuse and having a parent who is an alcoholic?

7. Describe the 4 components of emotions.

8. List the basic emotions and describe the timetable of development (according to Izard).

9. Describe how children's fears develop.

10. Define personality and temperament.

11. Describe the three traits that Buss and Plomin suggest constitute temperament.
    a.

    b.

    c.

12. Describe the components that Thomas and Chess identified as being part of temperament.

13. Describe the development of self-awareness.

14. Discuss the development of autonomy and separation-individuation.

15. Discuss the sources of self-efficacy and self-esteem.

KEY TERMS

In your own words, provide a definition for each of the following terms:

1. Attachment_____

2. Nonattached children_____

3. Insecurely attached_____

4. Separation anxiety_____

5. Stranger fear_____

6. Autistic phase_____

7. Symbiosis_____

8. Epinephrine_____

9. Personality_____

10. Temperament _____

11. Attention-deficit disorder _____

12. Separation-individuation _____

13. Self-reference _____

14. Self-efficacy _____

15. Self-esteem _____

APPLICATIONS

For each of the following, fill in the blank with one of the terms listed above.

1. Whenever Lenny's parents return home, he makes no effort to greet them, nor does he seem to care that they have come home. Lenny may be _____ to his parents.

2. According to Thomas and Chess, an infant who is on a regular schedule, is generally happy, and adjusts easily to new situations has an easy _____.

3. A child who thinks that he is not intelligent, that people don't like him, and in general, has a very negative impression of his own worth can be said to have low _____.

4. A child who runs to his father rather than a stranger when he is upset, and who wants to be held and talked to has developed a specific _____ to his father.

5. A person whom Emily has never met before comes into the room. Emily becomes very upset and clings to her mother. Emily is experiencing _____.

6. Eddie is excessively active, has great difficulty concentrating for any length of time in school and often jumps up out of his seat. Eddie may have _____.

7. According to Mahler, a 1 month old baby who believes that her mother exists to satisfy her basic needs is in the _____ phase.

8. According to Mahler, when an infant has established dependency on her mother, they are in the phase of _____.

9. Sam is very dependent on his parents and he cries a lot if they try to leave him. He won't tolerate being away from them at all, and always wants to be held. He may be _____ to his parents.

10. The hormone that is secreted by the adrenal glands that produces arousal is _____.

SELF-TEST MULTIPLE CHOICE QUESTIONS

Circle the best answer for each question.

1. Attachment relationships can
    a. provide the child with a sense of security.
    b. help children get along with others.
    c. become the basis for personality formation.
    d. all of the above

2. It is not generally a good idea to change caregivers often because
    a. infants can only become attached to one person so it is difficult to switch caregivers.
    b. infants can only distinguish between a few different individuals, not a lot of different people.
    c. stability of care is one of the most important elements in the maintenance of emotional security.
    d. all of the above

3. The sequence of developing attachments is from
    a. people in general to specific individuals.
    b. specific individuals to people in general.
    c. multiple attachments to single attachments.
    d. single attachments to multiple attachments.

4. Attachment behaviors are at their peak
    a. before 6 months of age.
    b. from 6 to 12 months of age.
    c. from 12 to 18 months of age.
    d. during the third year.

5. A child who constantly cries to be picked up and then continues to cry when he is held, and generally is very clingy to his parents may be
    a. nonattached.
    b. insecurely attached.
    c. precocious.
    d. multiply attached.

6. A child who is separated for a long time from her parents will likely go through which of the following sequences of phases?
   a. protest, despair, detachment
   b. despair, detachment, protest
   c. protest, detachment, despair
   d. despair, protest, detachment

7. Protest over temporary separations declines most sharply at around
   a. 1 year.
   b. 2 years.
   c. 3 years.
   d. 4 years.

8. Children do not generally exhibit a fear of strangers until after about
   a. 2 months.
   b. 4 months.
   c. 6 months.
   d. 2 years.

9. According to Erik Erikson, the "cornerstone of a vital personality" is the development of
   a. symbiosis.
   b. autonomy.
   c. independence.
   d. basic trust.

10. According to Mahler, the phase during which children establish dependency on their mother is the
    a. autistic phase.
    b. symbiotic phase.
    c. basic trust phase.
    d. autonomy phase.

11. In order for a child to develop a sense of trust and security, s/he must
    a. receive regular and adequate feedings.
    b. get sufficient sucking.
    c. experience physical contact and love.
    d. all of the above

12. As a young child, Chloe was frequently criticized by her parents and she never felt as if she could please them. As a result, Chloe may have a hard time developing
    a. a sense of autonomy.
    b. individuation.
    c. a sense of security and trust.
    d. a sense of reality.

13. Parents who are overprotective may create a child who has difficulty
    a. establishing autonomy.
    b. doing well in school.
    c. making friends.
    d. dealing with frustration.

14. Which of the following is not considered to be a function of emotions?
    a. They are adaptive and help ensure survival.
    b. They can indicate when physiological needs such as hunger should be met.
    c. They are a useful means of communication.
    d. They are powerful motivators of development.

15. Which of the following is not considered to be a basic emotion by Ekman?
    a. jealousy
    b. surprise
    c. disgust
    d. anger

16. According to Izard, which of the following emotions is present at birth?
    a. joy
    b. anger
    c. sadness
    d. disgust

17. The relatively consistent, basic disposition which is inherent in every person is referred to as
    a. personality.
    b. temperament.
    c. reactivity.
    d. emotionality.

18. Which of the following is not a trait of temperament as specified by Buss and Plomin?
    a. emotionality
    b. reactivity
    c. sociability
    d. activity

19. Enrico is a very active child with irregular schedules. He is prone to having temper tantrums when he doesn't get what he wants. According to Thomas and Chess, he would be classified as
    a. an easy child.
    b. a slow-to-warm-up child.
    c. a reactive child.
    d. a difficult child.

20. A measure of the point at which a child responds to a sound would be an example of a measure of Thomas and Chess' temperament component of
    a. rhythmicity.
    b. adaptability.
    c. sensory threshold.
    d. persistence.

21. Self-awareness means knowing that
    a. you are physically separate from other people.
    b. you are good or bad at sports.
    c. you are smarter or dumber than other people your age.
    d. you have arms and legs.

22. According to Erikson, the major psychosocial task between 1 and 2 years of age is the development of
    a. trust.
    b. dependency.
    c. symbiosis.
    d. autonomy.

23. According to Mahler, the period between 5 months and 3 years involves
    a. the development of trust.
    b. the development of separation-individuation.
    c. growing dependency needs.
    d. a fusing of personalities with the mother.

24. Greg did very poorly on a task, but when he is asked how well he will do it the next time, he says that he will do very well. Based on just this information, how old do you think Greg is?
    a. 1 year
    b. 3 years
    c. 8 years
    d. 12 years

25. According to Bandura, the <u>emotional</u> source of self-efficacy refers to
    a. if a child feels badly because he did not do well on a task.
    b. how the child feels in comparison to her peers.
    c. whether the child has been encouraged or discouraged.
    d. the level of physiological arousal which can affect judgements either positively or negatively.

ANSWER KEY

APPLICATIONS

1. nonattached
2. temperament
3. self-esteem
4. attachment
5. stranger fear
6. attention-deficit disorder
7. autistic
8. symbiosis
9. insecurely attached
10. epinephrine

MULTIPLE CHOICE

1. d (p. 232)
2. c (p. 233)
3. a (p. 234)
4. c (p. 234)
5. b (p. 235)
6. a (p. 236)
7. c (p. 238)
8. c (p. 238)

9. d (p. 241)
10. b (p. 241)
11. d (pp. 241-242)
12. c (p. 243)
13. a (p. 245)
14. b (p. 246)
15. a (p. 248)
16. d (p. 249)

17. b (p. 251)
18. b (p. 252)
19. d (p. 253)
20. c (p. 252)
21. a (p. 253)
22. d (p. 254)
23. b (p. 256)
24. b (p. 256)
25. d (p. 258)

Chapter 9

SOCIAL DEVELOPMENT

CHAPTER OUTLINE

I. Sociocultural Influences - Bronfenbrenner developed an ecological model for understanding social influences. The child is at the center of the model and is surrounded by systems of external influences, which have both positive and negative effects.

II. The Family and Socialization

    A. The Family's Role
        1. Socialization is the process by which persons learn the ways of society or social groups so that they can function within it or them.
        2. Children are socialized initially through the family through formal instruction, rewards and punishment, reciprocal parent-child interaction, and observational modeling.
        3. Different types of family structure and individual differences also affect socialization patterns.

    B. Parental Competence and Family Environment
        1. Parents' own psychological adjustment will affect how they raise their children.
        2. The quality of the marital relationship will affect children's development, but not all conflict is harmful.

    C. Patterns of Parenting
        1. Baumrind has identified three general styles of parenting: authoritarian, permissive and authoritative.
        2. According to Schaefer, successful parenting seems to show the maximum amount of love, and a balance between autonomy and control.

    D. Discipline
        1. Discipline is more effective if used within a loving relationship.
        2. It should be consistent and involve both rewards and punishments.

3. It is most effective if applied soon after the offense and takes into account the child's age.
   4. Discipline that inflicts pain, is too strict or too often applied is less effective.
   5. Extremes of either permissiveness or authoritarianism are counterproductive.
   6. Discipline that threatens the child's self-esteem should be avoided.

E. The Father's Role - Even when mothers work full-time, many fathers do not participate in household chores to the same extent as do mothers.

F. Sibling Relationships
   1. First-born and last-born children tend to receive special attention.
   2. The greater the number of children in a family, the less likely they will complete their education.
   3. Whether a sibling is a boy or a girl, and is older or younger, can have positive or negative influences.
   4. Sibling rivalry is quite common and varies in degree of severity.

G. Grandparents - A close relationship with a grandparent can be extremely positive for a child. However, sometimes conflicts may arise between parents and grandparents over child-rearing.

III. Nonnuclear Families

A. One-Parent Families
   1. One of the most important problems of the female-headed family is limited income. In the male-headed family, although financial problems aren't as severe, the most common complaints are about financial pressures, and not spending enough time with their children.
   2. The earlier a boy is separated from his father, the more affected he will be in his early years. Father absence may also affect the development of masculinity and educational attainment. Paternal absence for girls may make it more difficult to relate to males during adolescence. In some situations, it is not better to have the father present.

B. Divorce and Children
   1. Short-term reactions include a period of mourning and grief.
   2. Other common reactions are a sense of insecurity and anxiety, a tendency to blame themselves, a preoccupation with reconciliation, and feelings of anger and resentment.
   3. African-Americans are less likely than whites to remarry, but are more likely to adjust well and to have extensive support.

C. Stepfamilies - There may be problems with stepfamilies because stepparents have unrealistically high expectations, the roles are ill-defined, they must overcome stereotypes, stepparents have to deal with children who have been socialized by other parents, and they must deal with unresolved emotional issues and a network of complex kinship relationships.

    D. Foster Care - Foster care is becoming more prevalent, but is far from ideal.

    E. Adoptive Families - Many adoptive parents also have children of their own and often are related to the child. The benefits of "open adoptions" have been questioned.

IV. The Development of Peer Relationships

    A. Psychosocial Development - Children pass through four stages: autosociality, childhood heterosociality, homosociality, and adolescent and adult heterosociality.

    B. Infants and Toddlers - Infants interact with one another from about 5 months, primarily through looking and smiling. By 9 months, they will offer toys.

    C. Early Childhood
       1. Two-year-olds will play alongside one another and show some preferences in playmates.
       2. By three years, more friendly encounters occur, and aggressive behavior gradually declines.
       3. By four and five years, children share affection and objects, and will begin to form larger groups of playmates.
       4. Negative social skills could result from aggressive parental models, restrictive discipline, or parents' hostile reactions to provocation by the child.

    D. Middle Childhood
       1. Friendships become more important with age.
       2. Peer acceptance is predictive of later adjustment.
       3. There is a difference between children who are unliked, who tend to be loners, and children who are disliked, who tend to behave inappropriately.
       4. Children who are rejected report feelings of loneliness. Relations with parents can also affect loneliness.
       5. Gangs and clubs arise out of children's need to be independent from parents and to be with peers.
       6. Many authorities believe that preadolescents should avoid highly competitive activities.
       7. Some children can be very cruel.

    E. Social Cognition - the capacity to understand social relationships. Selman has described 5 stages in the development of social role taking.

V. Television as a Socializing Influence

    A. Viewing habits - Children between the ages of 6 and 11 spend about 26 hours per week watching television.

    B. Violence and Aggression - Research indicates that television violence is associated with increased aggressive behavior in children who watch it. Children model and imitate what they see, and they come to accept aggression as appropriate behavior.

    C. Family Interaction - Extensive viewing has been associated with a decrease in family interaction and communication.

    D. Cognitive Development - Heavy television viewing has been associated with lower school achievement of children from high, but not low, SES families.

    E. Commercials - Television advertising is influential in both positive and negative ways.

    F. Positive Effects - Some educational programs, like "Sesame Street," have been shown to have positive benefits. Television can promote good health and nutrition habits and prosocial behavior.

VI. The Development of Gender Roles

    A. Meaning - Gender refers to our biological sex, and gender roles are outward expressions of masculinity and femininity in social settings.

    B. Influences on Gender Roles
       1. Biological - Gender development is influenced by the chromosomal combination and the sex hormones.
       2. Cognitive - When the child recognizes that s/he is male or female, s/he strives to act consistently with the appropriate gender.
       3. Environmental - The child learns sex-typed behavior through rewards and punishment, indoctrination, observation of others, and modeling.

    C. Age and Gender-Role Development
       1. By 2 years, children are aware of boys and girls and sex-typed roles.
       2. By 3 years, children choose sex-typed toys though they are unaware why - this changes by 5 years.
       3. By age 7, children have developed a sense of gender constancy.

    D. Stereotypes - Although standards of maleness and femaleness are undergoing change, there is still much evidence of traditional stereotyping.

E. Androgyny - Androgynous persons are not sex-typed with respect to roles, although they are distinctly male or female in gender.

VII. Moral Development

A. Moral Judgement - According to Piaget, children move from a morality of constraint to a morality of cooperation; from heteronomy to autonomy in making moral judgements, and from the concept of expiatory punishment to punishment of reciprocity. Changes in moral judgements are related to cognitive growth and changes in social relationships.

B. Moral Behavior - Moral behavior is affected by moral motivation and moral inhibition. Over time, children come to depend more on internal than external factors when determining what is right and wrong.

LEARNING OBJECTIVES/ STUDY QUESTIONS

After reading this chapter, you should be able to:

1. Describe Bronfenbrenner's ecological model for understanding social influences.

2. Describe some of the different types of family structures.

3. Discuss how parental factors can affect a child's development.

4. Describe the 3 types of parenting styles identified by Baumrind.
   a.

   b.

   c.

5. Discuss the ways in which discipline can enhance or debilitate development.

6. Discuss the different ways in which siblings can affect development.

7. Discuss the positive and negative aspects of grandchild-grandparent relationships.

8. Discuss the different effects on boys and girls of living in single-parent households.

9. Describe some of the common ways in which children react to divorce.

10. Discuss some of the problems faced by stepparents.

11. Describe some of the factors involved in foster care and adoption.

12. Describe the four stages of psychosocial development.
    a.
    b.
    c.
    d.

13. Describe how peer relationships change from infancy to middle childhood.

14. List some of the characteristics of popular and unpopular children.

15. Describe Selman's five stages of the development of social role taking.
    a. Stage 0 -

    b. Stage 1 -

    c. Stage 2 -

    d. Stage 3 -

    e. Stage 4 -

16. Describe some of the positive and negative effects that are related to children's television watching.

17. Describe the biological, cognitive and environmental influences on gender roles.
    a. Biological -

    b. Cognitive -

    c. Environmental -

18. Discuss how younger children's moral reasoning differs from older children's moral reasoning, according to Piaget.

19. Discuss how moral behavior changes over development.

KEY TERMS I

In your own words, provide a definition for each of the following terms:

1. Microsystem_____

2. Mesosystem_____

3. Exosystem_____

4. Macrosystem_____

5. Socialization_____

6. Generational transmission_____

7. Single-parent family_____

8. Nuclear family_____

9. Extended family_____

10. Blended or reconstituted family_____

11. Stepfamily_____

12. Binuclear family_____

13. Communal family_____

14. Homosexual family_____

15. Cohabitating family_____

16. Sibling rivalry_____

17. Open adoption_____

APPLICATIONS I

For each of the following, fill in the blank with one of the terms listed above.

1. In Bronfenbrenner's ecological model, how parental discipline affects the child's behavior in the classroom would be considered part of the _____.

2. The process by which children learn how to fit in to their society is called _____.

3. Mario, recently divorced, marries Stephanie to form a _____ or _____ family.

4. Cultural values and beliefs are considered to be part of the _____ in Bronfenbrenner's model of social influences.

5. Natalie lives with her husband and their six children. This type of family is called a _____.

6. In Bronfenbrenner's ecological model, the child's friends would be included in the _____.

7. Children learn what is expected of them from their parents, who have learned it from their parents. This process of teaching is referred to as _____.

8. Nicole lives part of the time with her mother and stepfather and part of the time with her father. She is part of a _____ family.

9. Joan and Carol, with their two children, live together and are sexually and emotionally committed to one another. They are part of a _____ family.

10. A father is treated badly at work by his boss, and then he takes it out on his child when he gets home. In Bronfenbrenner's model of social influences, what took place at the father's work would be part of the _____.

KEY TERMS II

In your own words, provide a definition for each of the following terms:

1. Father hunger _____

2. Autosociality _____

3. Childhood heterosociality _____

4. Homosociality _____

5. Adolescent and adult heterosociality _____

6. Social cognition _____

7. Social role taking _____

8. Gender _____

9. Gender roles _____

10. Gender constancy _____

11. Gender stereotypes _____

12. Androgyny _____

13. Morality of constraint _____

14. Morality of cooperation _____

15. Objective judgements _____

16. Subjective judgements _____

17. Expiatory punishment _____

18. Punishment of reciprocity _____

19. Imminent justice _____

20. Equity _____

APPLICATIONS II

For each of the following, fill in the blank with one of the terms listed above.

1. Our outward expressions of masculinity and femininity are referred to as our _____.

2. The period of time in middle childhood during which children prefer to play with others of the same sex is referred to as _____.

3. The idea that little girls only like to play with dolls and other household items is a _____.

4. One month after his father left, Anthony, who was 2 years old at the time, began experiencing sleep disturbances. This syndrome is sometimes called _____.

5. A child who is assertive and independent, yet is also caring and sympathetic towards others may be described as _____.

6. A child who can figure out what other people think of her behavior is skilled at _____.

7. During infancy and toddlerhood, children prefer to play by themselves or alongside of another child, but not really with another child. This stage is referred to as _____.

8. A child who understands that she is a girl and will remain a girl, no matter how she dresses or behaves, understands _____.

9. Children initially believe that rules reflect parental authority and thus, they are inviolable, what Piaget called a morality of _____. Later, they come to understand that rules can be changed through mutual consent, what Piaget referred to as a morality of _____.

10. According to Piaget, punishment which is imposed by an authority figure is called _____, whereas punishment that is self-imposed is called _____.

SELF-TEST MULTIPLE CHOICE QUESTIONS

Circle the best answer for each question.

1. In Bronfenbrenner's model for understanding social influences, the system which is concerned with the relationships among the child's immediate contacts is the
    a. microsystem.
    b. mesosystem.
    c. exosystem.
    d. macrosystem.

2. The system that would include the differences between inner-city vs. suburban values and attitudes is the
    a. microsystem.
    b. mesosystem.
    c. exosystem.
    d. macrosystem.

3. Len and Linda, who are not married, have two children and are committed to each other. They are a
    a. binuclear family.
    b. blended family.
    c. nuclear family.
    d. cohabitating family.

4. The process of socialization in the family takes place through
    a. formal instruction.
    b. reciprocal parent-child interaction.
    c. observational modeling.
    d. all of the above

5. Research suggests that parents who are depressed may affect their children's behavior because depressed parents
    a. are less affectionate to each other.
    b. show less warmth to their children.
    c. tend to have more marital conflict.
    d. all of the above

6. A parent who uses strict, and often physical, punishment, and who has the attitude of "what I say goes" would most likely be classified as using which type of parenting style?
    a. authoritarian
    b. authoritative
    c. permissive
    d. normal

7. A parent who points out to her child that he should not hit another child because it might hurt that child is using which type of control technique?
   a. power-assertive discipline
   b. command
   c. other-oriented induction
   d. self-oriented induction

8. Research has suggested that the one parenting variable that is most related to children's adjustment is
   a. autonomy.
   b. love.
   c. control.
   d. permissiveness.

9. If you were to do a survey of students graduating from medical school, what would you expect to find given the findings of previous studies?
   a. First-born children would be overrepresented.
   b. Second-born children would be overrepresented.
   c. Middle children would be overrepresented.
   d. Youngest children would be overrepresented.

10. Older siblings
    a. are a positive influence on their younger siblings.
    b. are a positive influence on their younger siblings only if they are of the opposite sex.
    c. are a positive influence on their younger siblings only if they have a good relationship.
    d. are a negative influence on their younger siblings if they are of the opposite sex.

11. Girls who were raised without their fathers around, in comparison to girls who were raised by both of their parents, were more likely to
    a. have had problems dealing with the opposite sex when they were adolescents.
    b. have had sleep disturbances as young children.
    c. have had more problems when they were younger than when they were older.
    d. have shown masculine behaviors.

12. Newly formed stepfamilies may have problems adjusting to their new way of life because
    a. the stepparents may have unrealistically high expectations.
    b. the stepparents may be unsure of what their roles are.
    c. the stepparents may be faced with unresolved emotional issues from the prior marriage and divorce.
    d. all of the above

13. Which of the following is <u>not</u> true about open adoptions?
    a. The natural mother plays a role in choosing the adoptive parents.
    b. The natural mother may have contact with the child after the adoption has taken place.
    c. Open adoptions only occur between relatives.
    d. Open adoption is usually an expensive undertaking.

14. Autosociality refers to
    a. the stage during which infants and toddlers prefer to play by themselves.
    b. the stage during which young children try to assert their independence.
    c. the stage during which adolescents try to assert their independence.
    d. having a good time while riding in a car.

15. Children will often begin to play in groups by the time they are
    a. 1 year old.
    b. 18 months.
    c. 2 years.
    d. 3 years.

16. Joyce is a disliked child, while Maureen is an unliked child. Which of the following is the best statement regarding these two children?
    a. Joyce is probably very aggressive and obnoxious, while Maureen is more likely shy and withdrawn.
    b. Both Joyce and Maureen will probably be more accepted as they get older.
    c. Maureen is likely to become a disliked child as she gets older, while Joyce is more likely to be accepted later on.
    d. Joyce is likely to remain disliked, while Maureen may very well become accepted as she gets older.

17. Jodie is asked how another child will feel if he is given some milk to drink. She replies that the other child will be unhappy, based on the fact that she herself does not like milk. Jodie is at which of Selman's developmental stages?
    a. Egocentric undifferentiated stage
    b. Differentiated or subjective perspective-taking stage
    c. Self-reflective thinking or reciprocal perspective-taking stage
    d. Third-person or mutual perspective-taking stage

18. Gerald is told a story about how another child has received a gift that she didn't like. Gerald is also shown a picture of this child smiling after having opened the present. When Gerald is asked how he thinks the girl is feeling, he replies "happy," based on how she looks. Gerald is at which of Selman's developmental stages?
    a. Egocentric undifferentiated stage
    b. Differentiated or subjective perspective-taking stage
    c. Self-reflective thinking or reciprocal perspective-taking stage
    d. Third-person or mutual perspective-taking stage.

19. Research looking at the effects of watching violence on television has found that
    a. watching violence on television causes children to act aggressively.
    b. watching violence on television is related to increased aggressive behavior.
    c. watching cartoons leads to more aggressive behavior than watching real people behave violently on television.
    d. watching violence on television has no effects on children's behavior.

20. Research has found that extensive television watching leads to
    a. increased family interaction and communication.
    b. lower school achievement for children from low SES families.
    c. lower school achievement for children from higher SES families.
    d. all of the above

21. Our biological sex, either male or female, is referred to as
    a. gender.
    b. gender roles.
    c. gender constancy.
    d. sex-typing.

22. According to which of the following theories does sex-role identity begin to form when a child realizes that s/he is a boy or a girl, and then tries to act consistently with gender expectations?
    a. biological
    b. cognitive
    c. environmental
    d. psychoanalytic

23. A child who thinks that a boy who puts on a dress and plays with dolls will become a girl does not yet understand the concept of
    a. gender constancy.
    b. gender roles.
    c. gender stereotypes.
    d. gender identities.

24. An androgynous child is one who
    a. has neither masculine nor feminine characteristics.
    b. has both masculine and feminine characteristics.
    c. is either a boy who has only feminine characteristics or is a girl who has only masculine characteristics.
    d. has both male and female biological traits.

25. A child who believes that rules are set by authority figures and cannot be broken believes in a morality of
    a. restriction.
    b. authoritarianism.
    c. cooperation.
    d. constraint.

26. According to Piaget, children move from
    a. a morality of cooperation to a morality of constraint.
    b. subjective judgements to objective judgements.
    c. heteronomy to autonomy in making moral judgements.
    d. all of the above

27. A child who believes in a morality of constraint most likely also believes in the concept of
    a. punishment of reciprocity.
    b. cooperation.
    c. expiatory punishment.
    d. autonomy.

28. In response to his own conscience, John punished himself by not allowing himself to participate in some activity. In Piaget's terms, this type of punishment is
    a. expiatory punishment.
    b. punishment of reciprocity.
    c. preoperational punishment.
    d. heteronomous punishment.

29. A child who believes that bad behavior will naturally and inevitably be punished in some way believes in the concept of
    a. imminent justice.
    b. expiatory punishment.
    c. punishment of reciprocity.
    d. equity.

30. Research suggests that preschoolers
    a. think highly of children who tell the truth about a misdeed.
    b. think more negatively about a child who has been punished for a misdeed than they do about a child who has committed a misdeed but not been punished.
    c. do not appreciate that lying about a misdeed is bad.
    d. rely on internal rather than external factors when evaluating whether something is right or wrong.

ANSWER KEY

APPLICATIONS I

1. mesosystem
2. socialization
3. blended or reconstituted
4. macrosystem
5. nuclear family
6. microsystem
7. generational transmission
8. binuclear
9. homosexual
10. exosystem

APPLICATIONS II

1. gender roles
2. homosociality
3. gender stereotype
4. Father hunger
5. androgynous
6. social role-taking
7. autosociality
8. gender constancy
9. constraint; cooperation
10. expiatory punishment; punishment of reciprocity

MULTIPLE CHOICE

1. b (p. 266)
2. d (p. 266)
3. d (p. 269)
4. d (p. 269)
5. d (p. 272)
6. a (p. 273)
7. c (p. 275)
8. b (p. 275)
9. a (p. 279)
10. c (p. 281)
11. a (p. 287)
12. d (p. 290)
13. c (p. 294)
14. a (p. 295)
15. d (p. 296)
16. a (p. 300)
17. a (p. 303)
18. b (p. 303)
19. b (p. 305)
20. c (p. 306)
21. a (p. 308)
22. b (p. 308)
23. a (p. 310)
24. b (p. 311)
25. d (p. 312)
26. c (p. 312)
27. c (p. 313)
28. b (p. 313)
29. a (p. 313)
30. b (p. 314)

Chapter 10

PERSPECTIVES ON ADOLESCENT DEVELOPMENT

CHAPTER OUTLINE

I. The Meaning of Adolescence

    A. Adolescence - Adolescence is a period of growth beginning with puberty and ending at the beginning of adulthood.

    B. Puberty and Pubescence
       1. Puberty is the period at which a person reaches sexual maturity and becomes capable of having children.
       2. Pubescence describes the period during which physical changes relative to sexual maturation are taking place. Psychological and social changes take place as well.

    C. Maturity - Maturity is the age, state, or time of life at which a person is considered fully developed socially, intellectually, emotionally, physically, and spiritually.

    D. Juvenile - The word juvenile is a legal term describing an individual who is not accorded adult status in the eyes of the law.

II. Adolescence and Psychic Disequilibrium

    A. Storm and Stress - G. Stanley Hall described adolescence as a period of great storm and stress, the causes of which are biological. Researchers no longer believe that storm and stress are inevitable consequences of adolescence.

    B. Psychic Conflict - Anna Freud also characterized adolescence as a period of psychic disequilibrium, emotional conflict, and erratic behavior. Instinctual forces that have remained latent since early childhood reappear. The increasing demands of the id during adolescence create conflict with the superego, which the ego must try to resolve.

III. Adolescence and Identity Achievement

   A. Components of Identity - According to Erikson, the chief psychosocial task of adolescence is the achievement of identity. Identity has many components - sexual, social, physical, psychological, moral, ideological, and vocational.

   B. Psychosocial Moratorium - Erikson described a period of adolescence during which the individual may stand back, analyze, and experiment with various roles without assuming any one role.

   C. Identity Statuses
      1. Marcia formulated four identity statuses: identity achievement, moratorium, foreclosure, and identity diffusion. These do not always develop in exact sequence.
      2. Some researchers argue that women's identity formation deals with different issues such as intimacy and significant relationships.

   D. Ethnic Identity
      1. Ethnic identity involves positive ethnic attitudes, ethnic identity achievement, and ethnic behaviors.
      2. Members of ethnic groups participate in society through assimilation, integration, separation, or marginality.
      3. Gifted black students often report feeling as if they don't fit in.

IV. Adolescence and Developmental Tasks

   A. Meaning - Developmental tasks are the knowledge, attitudes, functions, and skills that individuals must acquire at certain points in their lives through physical maturation, personal effort, and social expectations.

   B. Eight Major Tasks:
      1. Accepting one's physique and using the body effectively
      2. Achieving emotional independence from parents and other adults
      3. Achieving a masculine or feminine social-sex role
      4. Achieving new and more mature relations with age-mates of both sexes
      5. Desiring and achieving socially responsible behavior
      6. Acquiring a set of values and an ethical system as a guide to behavior
      7. Preparing for an economic career
      8. Preparing for marriage and family life

V. Anthropologists' Views of Adolescence

   A. Developmental Continuity Versus Discontinuity - Anthropologists, such as Margaret Mead, emphasize continuity of development rather than the discontinuity of different stages. For example, Samoans never have to abruptly change their ways of acting or thinking as they move from childhood to adulthood. Rather than being submissive children who turn into dominant adults, Samoan children become continuously more dominant with age.

B. Cultural Influences - Anthropologists say that storm and stress during adolescence is not inevitable.

C. Generation Gap - Anthropologists deny the inevitability of a generation gap.

LEARNING OBJECTIVES/ STUDY QUESTIONS

After reading this chapter, you should be able to:

1. Discuss why G. Stanley Hall believes that adolescence is a time of "storm and stress."

2. Discuss Anna Freud's concept of psychic conflict in adolescence and how it can become problematic.

3. Discuss the formation of identity during adolescence, including the four identity statuses. How does it differ for women?

4. Define ethnic identity.

5. Discuss how members of ethnic groups participate in society. How do these options relate to self-esteem?

6. Discuss Havighurst's eight major psychosocial tasks to be accomplished during adolescence.
   a.

   b.

   c.

   d.

   e.

   f.

   g.

   h.

7. Discuss anthropologists' views of the continuity vs. discontinuity of development and how it differs from what is commonly held to be the case in Western cultures.

KEY TERMS

In your own words, provide a definition for each of the following terms:

1. Maturity _____

2. Puberty _____

3. Pubescence _____

4. Juvenile _____

5. Psychosocial moratorium _____

6. Identity achievement _____

7. Moratorium _____

8. Foreclosure _____

9. Identity diffusion _____

10. Acculturation _____

11. Developmental tasks _____

APPLICATIONS

For each of the following, fill in the blank with one of the terms listed above.

1. Bill has grown as much as he ever will; he can be said to have reached physical _____.

2. Developing mature relationships with people of both sexes is considered to be a
_____.

3. A 14-year-old who commits a crime would be treated in the eyes of the law as a _____ rather than as an adult.

4. The period of time between childhood and adulthood where an individual can spend some time exploring different roles without feeling pressured to know just what he or she wants is referred to as _____.

5. Donna has started menstruating and she is now capable of reproduction. She has reached _____.

6. Ron's parents are both lawyers. Ever since he was little, his parents told Ron that he too would become a lawyer. When Ron's 10th grade teacher asked him to write an essay on what he wanted to do with his life, without giving it any thought, Ron wrote about how he would become a lawyer. Ron's identity status would be _____.

7. For several years, Dana agonized over whether she should go to business school or a liberal arts school. After thinking through it very carefully, she decided to go to business school, and she was proud of her decision. Dana's identity status would be
_____.

SELF-TEST MULTIPLE CHOICE QUESTIONS

Circle the best answer for each question.

1. Adolescence is
    a. a period in development which everyone goes through for the same length of time.
    b. a period in development, the length of which differs for different individuals.
    c. practically non-existent in industrialized societies.
    d. a legal term, but not a psychological term.

2. The period during which physical changes relative to sexual maturation take place is called
    a. the maturational stage.
    b. the stage of sexual transformation.
    c. adolescent turmoil.
    d. pubescence.

3. According to G. Stanley Hall, puberty is a time of emotional upset and instability because
   a. the id is dominating the superego.
   b. psychic conflict is not being resolved.
   c. of the biological changes of puberty.
   d. of insufficient rites of passage.

4. According to Anna Freud, adolescents experience conflicting behaviors, such as oscillating between rebellion and conformity, because
   a. of sexual maturation which causes psychic disequilibrium.
   b. children are not cognitively prepared to understand the changes in their bodies.
   c. our cultures do not provide adequate rites of passage to adulthood.
   d. of hormonal imbalances.

5. According to Anna Freud, if the ego takes the side of the id over the superego, the individual will
   a. be overly constrained and polite.
   b. exhibit uninhibited gratification of desires.
   c. be overwhelmed by guilt and anxiety.
   d. have a very active conscience.

6. According to Erikson, the major psychosocial task of adolescence is the achievement of
   a. autonomy.
   b. industry.
   c. identity.
   d. generativity.

7. Which component of identity is likely to develop first?
   a. physical
   b. vocational
   c. moral
   d. ideological

8. Which of the following identity statuses is the least developmentally advanced?
   a. identity achievement
   b. moratorium
   c. foreclosure
   d. identity diffusion

9. Rajeesh moved to the U.S. from Pakistan several years ago with his family. He has adjusted well to his new school and has many new friends, many of whom are white. He also believes that it is important to know as much as possible about his own ethnic background. He is proud of his history, and follows the customs very closely. Which acculturation option has Rajeesh chosen?
   a. assimilation
   b. marginality
   c. separation
   d. integration

10. A child who is rebellious against her parents may be trying to resolve which developmental task as suggested by Havighurst?
    a. accepting one's physique and using the body effectively
    b. achieving emotional independence from parents and other adults
    c. desiring and achieving socially responsible behavior
    d. acquiring a set of values and an ethical system as a guide to behavior

11. An adolescent who has recently become very involved in working with conservation groups in her community may be working on which of Havighurst's developmental tasks?
    a. achieving a masculine or feminine social-sex role
    b. desiring and achieving socially responsible behavior
    c. acquiring a set of values and an ethical system as a guide to behavior
    d. preparing for an economic career

12. Some anthropologists, such as Margaret Mead, believe that
    a. development occurs in a number of qualitatively different stages.
    b. the period of adolescence is particularly pronounced in non-Western cultures.
    c. development is continuous and thus adolescence as a transitional stage is practically nonexistent.
    d. storm and stress during adolescence is inevitable.

ANSWER KEY

APPLICATIONS

1. maturity
2. developmental task
3. juvenile
4. psychosocial moratorium
5. puberty
6. foreclosure
7. identity achievement

MULTIPLE CHOICE

1. b (p. 326)
2. d (p. 326)
3. c (p. 328)
4. a (p. 328)
5. b (p. 329)
6. c (p. 329)
7. a (p. 330)
8. d (p. 331)
9. d (p. 332)
10. b (p. 334)
11. b (p. 334)
12. c (p. 335)

Chapter 11

PHYSICAL DEVELOPMENT

CHAPTER OUTLINE

I. The Endocrine Glands and Hypothalamus - An endocrine gland is a gland that secretes hormones internally.

    A. Pituitary Gland - The pituitary gland is located in the base of the brain and produces hormones that regulate growth.

    B. Gonads - The ovaries in the female secrete estrogens and progesterone. The testes in the male secrete testosterone.

    C. Adrenals and Hypothalamus -
      1. The adrenal glands are located just above the kidneys.
      2. The hypothalamus is a small area of the forebrain that regulates such functions as lactation, pregnancy, menstrual cycles, hormonal production, eating, and sexual response.

II. Maturation and Functions of Sex Organs

    A. Male - Important changes occur in the male sex organs during adolescence, the most important of which is the development of mature sperm cells.

    B. Female - Many changes occur in the female internal and external sex organs during adolescence.

    C. Menstruation
      1. The average age of menarche is 12-13 years, although this varies considerably. The time of ovulation is ordinarily about 14 days before the onset of the next menstrual period.
      2. Heredity and ethnic factors, nutrition, medical care, and stress can all affect the timing of sexual maturation.

III. Physical Growth and Development

   A. Development of Secondary Sexual Characteristics - Sexual maturation at puberty also includes the development of secondary sexual characteristics.
   B. Growth in Height and Weight - A growth spurt in height begins in early adolescence and is accompanied by an increase in weight and changes in body proportions.

IV. Early and Late Maturation

   A. Early-Maturing Boys - Early-maturing boys have both athletic and social advantages.

   B. Early-Maturing Girls - Girls who mature early are at a disadvantage during the elementary school years, but may be at an advantage during junior and senior high school.

   C. Late-Maturing Boys - Late-maturing boys suffer a number of social disadvantages and may develop feelings of inferiority as a result.

   D. Late-Maturing Girls - Late-maturing girls of junior or senior high school age are likely to be socially handicapped.

V. Body Image and Psychological Impact

   A. Physical Attractiveness - Attractiveness affects the adolescent's positive self-esteem and social acceptance.

   B. Concepts of the Ideal - Adolescents are influenced by the concepts of the ideal build that are accepted by our culture. It is partly because of our obsession with slimness that anorexia nervosa and bulimia develop among adolescents.

VI. Sex Education of Adolescents

   A. Goals - Some of the goals of sex education include: gaining knowledge about bodily changes and basic facts about human reproduction, developing sexual health, preventing unwanted pregnancies and sexually transmitted diseases.

   B. The Parents' Role - Some parents do not do a good job discussing sex education with their children. Some may be uncomfortable or uninformed. Others may want to avoid experimentation.

   C. The School's Role
      1. Many schools are taking the responsibility for teaching sex education in an attempt to reach as many children as possible.

2. Surveys suggest there is a gap between teachers' beliefs about what and when topics should be covered and when and if they actually are covered.
3. Some programs that emphasize abstinence have been shown to delay initiation of sexual activity.

VII. Nutrition and Weight

   A. Caloric Requirements - During the period of rapid growth, adolescents need greater quantities of food to take care of bodily requirements.

   B. Importance of Nutrition - Health maintenance depends partly on proper eating habits.

   C. Deficiencies - Many adolescents have diets that are deficient in thiamine and riboflavin, other vitamins, calcium, iron, and/or protein.

   D. Overweight and Underweight - Being overweight or underweight affects the adolescent's emotional adjustment, self-esteem, and social relationships.

   E. Anorexia Nervosa - Anorexia nervosa is a life-threatening emotional disorder characterized by an obsession with being slender.

   F. Bulimia - Bulimia is a binge-purge syndrome.

LEARNING OBJECTIVES/ STUDY QUESTIONS

After reading this chapter, you should be able to:

1. Discuss the functions of the pituitary gland, the adrenal glands, the gonads and the hypothalamus.
   a. the pituitary gland -

   b. the adrenal glands -

   c. the gonads -

   d. the hypothalamus -

2. Discuss the maturation of the male and female sex organs.
   a. male -

   b. female -

3. Discuss some of the factors that affect the timing of sexual maturation.

4. Discuss the development of the secondary sexual characteristics.

5. Discuss the factors which affect growth in height and weight.

6. Describe the psychological effects of early or late maturation for boys and girls.

7. Discuss the impact of physical attractiveness on adolescents' psychological development.

8. What are the goals of sex education? What roles do parents and the schools play?

9. Describe anorexia nervosa and bulimia.

KEY TERMS I

In your own words, provide a definition for each of the following terms.

1. Endocrine glands _____

2. Hormones _____

3. Pituitary gland _____

4. Gonadotropic hormones _____

5. Follicle-stimulating hormone (FSH) _____

6. Luteinizing hormone (LH) _____

7. Human growth hormone (HGH) _____

8. Prolactin _____

9. Gonads _____

10. Estrogens _____

11. Progesterone _____

12. Corpus luteum_____

13. Adrenal glands_____

14. Hypothalamus_____

15. GnRH (gonadotropin-releasing hormone)_____
_____

APPLICATIONS I

For each of the following, fill in the blank with one of the terms listed above.

1. The _____ are located just above the kidneys; they secrete adrenalin as well as androgens and estrogens in both men and women.

2. The _____ hormone stimulates development of the ovum, as well as estrogen and progesterone in females, and sperm and testosterone in males.

3. _____ are a group of feminizing hormones produced by the ovaries, and to some extent by the adrenal glands, in both males and females.

4. The biochemical substances that are secreted into the bloodstream by the endocrine glands to tell the other cells what to do are called _____.

5. The secretion of milk by a nursing mother is stimulated by _____.

6. Eating, drinking, and sexual responses are controlled by the area in the brain called the _____.

7. The master gland of the body that produces hormones that regulate growth is the _____.

8. The testes and the ovaries are referred to as the _____.

9. The pituitary hormone that regulates overall body growth is _____.

10. The hormone that controls the production and release of FSH and LH from the pituitary is _____ hormone.

KEY TERMS II

In your own words, provide a definition for each of the following terms:

1. Anabolic steroids _____

2. Penis _____

3. Scrotum _____

4. Testes _____

5. Prostate gland _____

6. Seminal vesicles _____

7. Epididymis _____

8. Cowper's glands _____

9. Urethra _____

10. Vas deferens _____

11. Nocturnal emissions _____
_____

APPLICATIONS II

For each of the following, fill in the blank with one of the terms listed above.

1. The _____ is a system of ducts, running from the testes to the vas deferens, in which sperm cells ripen and are stored.

2. In some instances, conception can occur even if the male has withdrawn his penis prior to ejaculation because the alkaline fluid secreted by the _____ sometimes contains sperm.

3. Some athletes, in order to build up their bulk, take _____ that sometimes can have deleterious effects.

4. The _____ doubles in length and girth during adolescence.

5. An adolescent boy who has an erotic dream that culminates in an orgasm has experienced what is referred to as a _____.

6. The pouch of skin that contains the testes is the _____.

7. The _____ gland secretes a portion of the seminal fluid.

8. The _____ is a tube that carries urine from the bladder to the outside; in males, it also carries semen.

9. The tubes running from the epididymis to the urethra that carry semen and sperm make up the _____.

10. The twin glands that secrete fluid into the vas deferens to enhance sperm viability are the _____.

KEY TERMS III

In your own words, provide a definition for each of the following terms:

1. Vagina _____

2. Fallopian tubes _____

3. Uterus _____

4. Ovaries _____

5. Clitoris _____

6. Labia majora _____

7. Labia minora _____

8. Hymen _____

9. Bartholin's glands _____

10. Menarche _____

11. Amenorrhea _____

12. Secondary sexual characteristics _____

13. Primary sexual characteristics _____

14. Precocious puberty _____

15. Anorexia nervosa_____

16. Bulimia_____

17. Acne_____

APPLICATIONS III

For each of the following, fill in the blank with one of the terms listed above.

1. A girl who gives birth to a child when she is only 8 years old must have gone through _____.

2. The large lips of tissue on either side of the vaginal opening are the _____.

3. The womb in which the baby grows and develops is the _____.

4. In girls, _____ usually does not occur until maximum growth rates in height and weight have been achieved.

5. The tissue that partly covers the vaginal opening is called the _____.

6. An adolescent who is obsessed with being thin, and has lost an excessive amount of weight for her body size may have _____.

7. The adolescent boy's changing voice and growth of facial and pubic hair are considered to be _____.

8. The glands on either side of the vaginal opening that secrete fluid during sexual arousal are the _____.

9. The tubes that transport the ova from the ovaries to the uterus are the _____.

10. A woman who is a long-distance runner may experience _____.

SELF-TEST MULTIPLE CHOICE QUESTIONS

Circle the best answer for each question.

1. The gland that is referred to as the master gland, and which produces hormones that regulate growth is the
   a. adrenal gland.
   b. Cowper's gland.
   c. prostate gland.
   d. pituitary gland.

2. Which of the following hormones affects the secretion of milk by the mammary glands of the breast?
   a. follicle-stimulating hormone
   b. luteinizing hormone
   c. prolactin
   d. human growth hormone

3. As the testes mature in the male, _____ production increases dramatically, whereas the level of the _____ increase only slightly.
   a. estrogens; progesterone
   b. testosterone; estrogens
   c. estrogens; testosterone
   d. progesterone; estrogens

4. The development of male or female characteristics is partly determined by
   a. the presence or absence of male vs. female hormones.
   b. the ratio of the levels of male to female hormones.
   c. the presence or absence of estrogen.
   d. the presence or absence of androgens.

5. In the male, the adrenal glands produce
   a. both estrogens and androgens, but greater amounts of androgens.
   b. both estrogens and androgens, but greater amounts of estrogens.
   c. equal amounts of estrogens and androgens.
   d. only androgens.

6. If the hypothalamus could be electrically stimulated, what might occur?
   a. feelings of hunger and thirst
   b. sexual feelings and thoughts
   c. production of hormones
   d. all of the above

7. Where are the sperm stored?
   a. the testes
   b. the vas deferens
   c. the epididymus
   d. the urethra

8. The inner walls of the vagina change their secretion from the _____ reaction of childhood to the _____ reaction in adolescence.
   a. alkaline; acid
   b. acid; alkaline

9. The follicles usually produce a mature ovum about
   a. once every two hours.
   b. once every 14 days.
   c. once every 28 days.
   d. once a year.

10. Which of the following may delay menarche?
    a. good nutrition
    b. vigorous exercise
    c. an increase in body fat
    d. all of the above

11. The time of ovulation is usually
    a. about 14 days before the onset of the next menstrual period.
    b. about 14 days after the onset of the menstrual period.
    c. during the menstrual period.
    d. about 7 days after the onset of the menstrual period.

12. Jeff has noticed that his voice is beginning to change, the shape of his body is changing, and he is growing pubic and facial hair. What he is noticing is the development of
    a. primary sexual characteristics.
    b. secondary sexual characteristics.
    c. deciduous sexual characteristics.
    d. latent sexual characteristics.

13. The average girl matures about
    a. the same time as the average boy.
    b. 1 year before the average boy.
    c. 2 years after the average boy.
    d. 2 years before the average boy.

14. The most important environmental factor in determining the height of a child is
    a. heredity.
    b. exercise.
    c. nutrition.
    d. none of the above; height is only determined by genetic factors.

15. Which of the following adolescents is most likely to have high self-esteem?
    a. an early-maturing boy
    b. an early-maturing girl
    c. a late-maturing boy
    d. a late-maturing girl

16. Early-maturing girls
    a. are more likely to be at a social advantage than early maturing boys when they are in elementary school.
    b. are likely to start dating later than most.
    c. are at a disadvantage in elementary school, but in junior and senior high school, may be at a social advantage.
    d. are less likely than late-maturing girls to have parents that worry about them and try to restrain their activities.

17. Marlene is thought of as very attractive, while Brooke is considered to be fairly plain. If other individuals were asked to rate them on a number of personality traits,
    a. Brooke would likely be considered to be more intelligent, but Marlene would be considered to be more friendly.
    b. Marlene would likely be considered to be more intelligent, but Brooke would be considered to be more friendly.
    c. Marlene would likely be considered to be more intelligent and more friendly.
    d. Brooke would likely be considered to be more intelligent and more friendly.

18. Acne may be formed because of overactive
    a. sebaceous glands.
    b. apocrine sweat glands.
    c. merocrine sweat glands.
    d. Bartholin's glands.

19. Research on teaching sexual abstinence suggests that
    a. it is more effective for males than females.
    b. programs aimed at junior high school students may be more effective than programs for high school students.
    c. programs are more effectively taught in suburban schools than in urban schools.
    d. females rate the programs more negatively than do males.

20. According to which theory is anorexia nervosa caused by a sexual conflict in which the individual is unwilling to accept adult sexuality, fears sexual intimacy, and thus, tries to delay sexual development?
    a. Psychobiologic regression theory
    b. Psychosexual theory
    c. Family systems theory
    d. Social theory

21. A person who goes on eating binges and then engages in purging and vigorous exercise may
    a. have anorexia nervosa.
    b. have a healthy attitude about her body.
    c. be considered healthy, but only if her body weight is normal.
    d. have bulimia.

ANSWER KEY

APPLICATIONS I

1. adrenal glands
2. luteinizing
3. Estrogens
4. hormones
5. prolactin
6. hypothalamus
7. pituitary gland
8. gonads
9. human growth hormone
10. gonadotropin-releasing

APPLICATIONS II

1. epididymus
2. Cowper's gland
3. anabolic steroids
4. penis
5. nocturnal emission
6. scrotum
7. prostate
8. urethra
9. vas deferens
10. seminal vesicles

APPLICATIONS III

1. precocious puberty
2. labia majora
3. uterus
4. menarche
5. hymen
6. anorexia nervosa
7. secondary sexual characteristics
8. Bartholin's glands
9. Fallopian tubes
10. amenorrhea

MULTIPLE CHOICE

1. d (p. 342)	8. a (p. 346)	15. a (p. 351)
2. c (p. 342)	9. c (p. 347)	16. c (p. 351)
3. b (p. 343)	10. b (p. 347)	17. c (p. 354)
4. b (p. 343)	11. a (p. 347)	18. a (p. 356)
5. a (p. 344)	12. b (p. 349)	19. b (p. 358)
6. d (p. 344)	13. d (p. 349)	20. b (p. 362)
7. c (p. 345)	14. c (p. 349)	21. d (p. 362)

Chapter 12

COGNITIVE DEVELOPMENT

CHAPTER OUTLINE

I. Formal Operational Thought

    A. Characteristics
       1. In formal operations, one can reason, systematize ideas, construct theories and test them.
       2. The fundamental property of adolescent thought is being able to maneuver between reality and possibility.
       3. Formal operational adolescents can use a set of symbols of symbols.
       4. Adolescents can orient themselves toward what is abstract and not immediately present.
       5. Thus, formal operational thinking involves introspection, abstract thinking, logical thinking, and hypothetical reasoning.

    B. Effects on Personality and Behavior
       1. Idealism - The ability to distinguish the possible from the real allows them to imagine ideal circumstances and to focus on long-term implications.
       2. Discrepancy - Early adolescents can formulate general principles but have difficulty with specific practices.
       3. Self-consciousness and egocentrism - The capacity to think about their own thoughts makes adolescents become acutely aware of themselves.
       4. Conformity - Adolescents have a greater potential for creativity but in reality they are less creative because of pressures to conform.
       5. Decentering and a life plan - True integration into society comes when the adolescent begins to affirm a life plan and adopt a social role.

    C. Critique of Piaget's Formal Operational Stage
       1. Ages and percentages - There is a great deal of variability in the age at which formal operations is achieved, and some individuals may never achieve it.
       2. Test level - The percentages of people reaching formal operations is dependent on the measures used.
       3. Cross-cultural studies - Formal thought is more dependent on social

experience than is sensorimotor or concrete thought.

    D. Adolescent Education and Formal Operational Thought - According to Piaget, the two goals of education are to create individuals who are capable of doing new things, and to form minds which can be critical.

    E. Problem-Finding Stage - There may be a fifth stage of development, called a problem-finding stage, during which the ability to discover problems develops.

II. Scholastic Aptitude

    A. SAT -
      1. The Scholastic Aptitude Test is one of the most widely used tests in the U.S. It is supposed to measure basic abilities; however, some research has indicated that coaching and cramming can improve scores.
      2. Some authorities have suggested that achievement tests would be a better way of predicting college success than SATs.
      3. Verbal scores have declined since 1986, while math scores have declined for males, but risen slightly for females.

    B. Proposed Revisions of the SAT - The greatest proposed changes are in the math section in which there is greater emphasis on critical thinking and "real life" problem solving.

    C. ACT - The ACT Assessment Program is the second most widely used college admissions test; this test emphasizes achievement in specific subjects.

III. School

    A. Trends in American Education
      1. During the past 50 years, the emphasis in schools has switched from a traditionalist position, in which the basics are emphasized, to a progressive position, in which the purpose of education is to prepare students for all of life, back to a more traditionalist position.
      2. Traditionalism was the dominant emphasis in American schools until the Depression. After Sputnik, schools became more concerned with math and sciences. Then, in the 1960's and 1970's, the emphasis switched again to place emphasis on relevance to the real world. More recently, emphasis is again being placed on the basics.

    B. Types of High Schools - A student is more likely to get a superior education at a private school than at a public school. Private schools differ from public schools in terms of educational characteristics and students' family backgrounds.

C. Tracking - The technique of tracking has been criticized for providing greater opportunities for privileged students, but not for less privileged students.

D. Dropouts -
1. Most dropouts occur during the high school years, with a greater percentage of blacks than whites leaving school, largely due to differences in socioeconomic status.
2. There are a number of factors that correlate with early school withdrawal: family relationships, pregnancy and marriage, employment and money, social adjustment and peer associations, scholastic factors, apathy, misconduct, and personality characteristics.
3. Some students drop out because they feel alienated.
4. Intervention programs can reduce the dropout rate considerably.

LEARNING OBJECTIVES/ STUDY QUESTIONS

After reading this chapter, you should be able to:

1. Describe the characteristics of Piaget's formal operational stage.

2. Describe some of the effects of formal operational thought on adolescents' personalities.

3. Discuss the effects of maturation and of culture on the achievement of formal operational thought.

4. Describe the fifth stage of cognitive development.

5. Discuss the arguments for and against the use of SATs as a predictor of success in college.

6. Describe the changes in the American education system over the past 50 years in terms of whether a progressive approach or a more traditionalist approach has been emphasized.

7. Discuss some of the differences between public and private high schools.

8. Discuss some of the factors that are associated with dropping out of high school early.

KEY TERMS

In your own words, provide a definition for each of the following terms:

1. Sociocentrism _____
_____

2. Personal fable _____
_____

3. Problem-finding stage _____
_____

4. Scholastic Aptitude Test (SAT) _____
_____

5. ACT Assessment Program (American College Testing program) ____
_____

6. Progressives _____
_____

7. Traditionalists _____
_____

APPLICATIONS

For each of the following, fill in the blank with one of the terms listed above.

1. If a college admissions board were interested in assessing prospective students' general abilities, they would likely require that the students take the _____.

2. An educator who believes that the purpose of school is to teach students math and science and how to read and write is a _____.

3. An adolescent who is concerned about the fate of society has shifted from egocentrism to _____.

4. The ability to discover problems and raise general questions about issues that are not well-defined is part of the proposed fifth stage of cognitive development, the _____ stage.

5. An adolescent's belief that she will not get into a car accident if she drives drunk would be termed _____ by Elkind.

6. An educator who believes that the schools are responsible for teaching students about facets of life such as sex education and drug education is a _____.

7. An educator who is interested in assessing a child's knowledge of a specific subject such as Social Studies would most likely want the student to take the _____.

SELF-TEST MULTIPLE CHOICE QUESTIONS

Circle the best answer for each question.

1. Piaget's final stage of cognitive development is referred to as
    a. the preoperational stage.
    b. the concrete operational stage.
    c. the formal operational stage.
    d. the problem-finding stage.

2. Piaget found in his experiment in which adolescents were to determine what affected the oscillatory speed of a pendulum that adolescents showed which of the following characteristics?
    a. The adolescents set about their investigation in a haphazard fashion.
    b. They recorded the results accurately and objectively.
    c. They had difficulty drawing conclusions from the results.
    d. all of the above

3. Which of the following would indicate that an adolescent had reached the formal operational stage?
    a. sticking with an initial opinion even in the face of contradictory evidence
    b. being able to make logical classifications
    c. being able to derive a proposition from several variables
    d. all of the above

4. The ability to use a set of symbols to represent other symbols allows an individual to
    a. understand algebraic signs.
    b. learn a second language.
    c. draw a cartoon.
    d. all of the above

5. According to Piaget, which of the following characterizes formal thinking?
    a. introspection
    b. abstract thinking
    c. hypothetical reasoning
    d. all of the above

6. It has been suggested that adolescents have a particularly idealistic view of how the world should be because of their ability to
   a. distinguish what is real from what is possible.
   b. be flexible in their thinking.
   c. use deductive reasoning.
   d. use inductive reasoning.

7. An adolescent who preaches about how animals should have the same rights as humans, but yet wears leather and eats meat, may
   a. believe that the practice of ideals is more important than the general principle.
   b. understand the general principle involved, but may not realize that specific practices should be done to support the ideal.
   c. not understand the abstractions involved in idealistic thinking.
   d. believe that others are more moral.

8. An adolescent's egocentrism involves
   a. believing that other people are as concerned with his appearance and behavior as he is.
   b. not understanding that another person sees things from a different line of vision.
   c. believing that other people have the same thoughts as she does.
   d. all of the above

9. Are adolescents more or less creative than younger children?
   a. Adolescents are too concerned with their own appearance to express much creativity.
   b. They are less creative because of their emphasis on logical thinking.
   c. Most adolescents tend to express more creativity as they get older.
   d. Adolescents tend to have more creative capabilities, but are less likely to express them because of increasing pressures to conform.

10. One researcher tested a group of adolescents and found that about 40% of them had reached the formal operational stage, while another researcher found that considerably less had reached this stage. These researchers may have found differing results because
    a. they may have used measures which differed in their criteria for formal thinking.
    b. they may have tested groups with different backgrounds in terms of SES.
    c. they may have tested different age groups.
    d. all of the above are possible

11. If a researcher tested a group of 11-year-olds and found that the children with the highest IQ's did just as poorly as the other children on tests of formal operational thinking, this would indicate that
    a. intelligence is determined by maturation.
    b. intelligence is determined by the environment.
    c. formal operational thinking is related to levels of maturation.
    d. formal operational thinking is induced by environmental stimulation at any age.

12. According to Piaget, the principle goal of education is to
    a. teach children to understand everything that is offered to them.
    b. create individuals who can be creative and do new things.
    c. teach children to use the materials that are presented to them.
    d. teach children to repeat what others have successfully accomplished in the past.

13. Some researchers have suggested that there is a stage beyond formal operations that some people reach that involves the ability to
    a. solve complex problems that are presented to them.
    b. be critical of material that is presented, rather than just accepting it as truth.
    c. reason abstractly and think of future possibilities.
    d. discover and describe problems that have not already been delineated.

14. The SAT was designed to measure
    a. the general level of intelligence.
    b. specific knowledge that has been acquired in certain subject areas.
    c. the ability to engage in formal operational thinking and problem-solving.
    d. basic abilities that are acquired over a lifetime.

15. Since 1986, average verbal scores on the SAT have _____, and math scores have _____ for females.
    a. increased; decreased
    b. decreased; increased
    c. decreased; remained the same
    d. remained the same; decreased

16. Proposed changes of the SAT include
    a. changing the scoring system.
    b. making a written essay mandatory.
    c. an emphasis on "real life" problem solving.
    d. banning calculators.

17. Which of the following subjects is not included as an Academic Test on the ACT Assessment Program?
    a. science reasoning
    b. foreign languages
    c. reading
    d. math

18. Those who argue that the child's purpose for attending school is to learn math, science, history, and languages are considered
    a. traditionalists.
    b. authoritarian.
    c. authoritative.
    d. progressives.

19. An educator who argues that children should be enrolled in sex education classes because they are not getting the knowledge that they need at home would be considered to be
    a. a traditionalist.
    b. a progressive.
    c. a reformist.
    d. a revolutionist.

20. In the 1950's, after the Soviet Union launched the first space satellite, the American school system
    a. took on a more progressive approach to schooling.
    b. revamped their system, particularly emphasizing revision of the math and sciences curricula.
    c. placed more emphasis on "hands on" experience.
    d. lowered graduation requirements so that more individuals could go on to college.

21. Research suggests that a student is more likely to receive a superior education at which type of high school?
    a. a Catholic school
    b. a private boarding school
    c. a private non-boarding school
    d. a private school in comparison to a public high school

22. The process of tracking
    a. groups together students of differing abilities.
    b. usually helps both privileged students and less privileged students.
    c. may benefit privileged students but not less privileged students.
    d. creates heterogeneous groups that work less efficiently than do homogeneous groups.

23. Adolescents who drop out of school are more likely to
    a. have feelings of hostility and resentment.
    b. experience excessive fear and anxiety.
    c. have feelings of inferiority.
    d. all of the above

24. Black youths are more likely than white youths to drop out of school because
    a. they are more likely to come from a lower SES background.
    b. black families have a different value system than white families.
    c. black youths are more influenced by peer pressures.
    d. none of the above

25. Which of the following is <u>not</u> considered to be a predictor of dropping out of school?
    a. poor relationships with parents
    b. being placed in vocational courses rather than academic courses
    c. negative influence of peers
    d. misbehavior in school

ANSWER KEY

APPLICATIONS

1. Scholastic Aptitude Test
2. traditionalist
3. sociocentrism
4. problem-finding
5. personal fable
6. progressive
7. ACT Assessment Program

MULTIPLE CHOICE

1. c (p. 370)	9. d (p. 374)	17. b (p. 379)
2. b (p. 370)	10. d (p. 376)	18. a (p. 380)
3. c (p. 371)	11. c (p. 376)	19. b (p. 380)
4. a (p. 371)	12. b (p. 377)	20. b (p. 381)
5. d (p. 372)	13. d (p. 377)	21. d (p. 385)
6. a (p. 372)	14. d (p. 377)	22. c (p. 385)
7. b (p. 372)	15. b (p. 378)	23. d (p. 389)
8. a (p. 374)	16. c (p. 379)	24. a (p. 390)
		25. b (p. 391)

Chapter 13

EMOTIONAL DEVELOPMENT

CHAPTER OUTLINE

I. Adolescent's Emotions

    A. The Components of Emotions - Emotions are states of consciousness that are accompanied by physiological arousal and which result in behavioral responses.

    B. Joyous States - Whether or not children are joyous, happy, and loving will depend on the events around them and interactions with people. By the time that children reach adolescence, they already exhibit well-developed patterns of emotional responses.

    C. Inhibitory States
       1. Fear - Generally, as children grow they lose some of their fears of material things and natural phenomena, but develop more fears relating to the self, and involving social relationships.
       2. Phobias - A phobia is an irrational fear that exceeds normal proportions and has no basis in reality.
       3. Worry and anxiety - These emotions are closely allied to fear, but they may arise from imagined unpleasant situations as well as from real causes.
       4. Some adolescents grow up in a worry-free environment; others grow up under conditions that cause constant worry and tension.
       5. Generalized anxiety disorder is when anxiety becomes so pervasive and tenacious that it interferes with normal functioning.

    D. Hostile States
       1. Anger - Anger in adolescence has many causes, such as restrictions on social life, attacks on their ego or status, criticism, shaming, rejection, the actions of others, or their own ineptitude. Men tend to express their anger, whereas women are socialized to inhibit their anger.
       2. Hatred - Hatred can be a more serious emotion than anger because it can persist over a longer period of time, is difficult to suppress, and may be expressed through words or actions in violent ways.

II. Self-Concept and Self-Esteem

   A. Definitions
      1. Self-concept is the view or impression that people have of themselves which develops over a period of many years. The self becomes increasingly differentiated with age.
      2. Self-esteem is the value individuals place on the selves they perceive. Research suggests that for girls, issues of health and physical development and home and family are related to self-esteem.

   B. Correlations - Self-esteem is related to social adjustment, emotional well-being, achievement, the goals that one sets, and acting out behavior, such as juvenile delinquency.

   C. Parental Roles in Development - The quality of parent-adolescent relationships, the type of parental control, the atmosphere of the home, and parents' own self-esteem are all related to an adolescents' self-esteem and self-concept.

   D. Socioeconomic Variables - The effects of SES on self-esteem are very variable.

   E. Racial Considerations - Self-esteem among blacks has risen along with racial pride. The self-esteem of blacks depends partly on the extent to which they have been exposed to white prejudices.

   F. Short-Term and Longitudinal Changes - Self-esteem is lowest at around 12 years of age, and gradually stabilizes during adolescence.

   G. Perfectionism - Neurotic perfectionists pursue excellence to an unhealthy extreme and are plagued by self-criticism, low self-worth, and stress.

III. Emotions and Behavioral Problems

   A. Drug Abuse
      1. The most frequently used drugs in the U.S. are alcohol, tobacco, and marijuana.
      2. A physical addiction is a physical dependency. Psychological dependency is when there is a persistent, emotional need for a drug.
      3. Youth are trying drugs at younger ages.
      4. Drug use may be divided into 5 patterns: social-recreational use, experimental use, circumstantial-situational use, intensified drug use, and compulsive use.
      5. There is a correlation between drug addiction and dependency and disturbed family relationships. Other correlates include: disturbed peer relationships, loneliness, rebelliousness, depression, and more frequent and unprotected sex.

6. Chronic alcoholism is characterized by compulsive drinking so that there is no voluntary control over the amount consumed.

B. Delinquency
1. In 1990, of all persons arrested, 16% were juveniles.
2. Psychological causes of delinquency include emotional and personality factors.
3. Sociological causes include family background influences, SES, neighborhood and community influences, peer group involvement, affluence and hedonistic values, violence in our culture, cultural change and unrest, drinking and drug usage, and school performance.
4. Biological causes may play a role in delinquency.
5. Many young people who kill were themselves abused.
6. Much juvenile crime is drug-related.
7. Parricide, the killing of a parent, is often in response to abuse.
8. One way to prevent delinquency is to identify who is at risk for getting into trouble and planning intervention programs.

C. Running Away
1. Over 700,000 youths, ages 10-17, run away from home annually.
2. Children who run away from home can be classified according to the degree of conflict with parents as runaway explorers, social pleasure seekers, runaway manipulators, runaway retreatists, and endangered runaways.
3. The simplest classification of runaways is to divide them into two groups: the running from and the running to groups.
4. The National Runaway Youth Program has promoted nation-wide assistance to youths who are vulnerable to exploitation and to dangerous encounters.

LEARNING OBJECTIVES/ STUDY QUESTIONS

After reading this chapter, you should be able to:

1. Discuss why emotions are important.

2. Describe the three categories of emotions:
   a. joyous states -

   b. inhibitory states -

c. hostile states -

3. Describe the four categories of sources of fear.
   a.

   b.

   c.

   d.

4. Describe phobias, anxiety and worry in adolescence.

5. Describe the meaning of self and the importance of realistic self-concepts.

6. Describe the meaning of self-esteem and its importance.

7. Discuss the factors that are important to the development of a positive self-concept.

8. Discuss the stability or change of self-concepts during adolescence.

9. Discuss aspects of drug abuse, such as which drugs are abused, physical addiction and psychological dependency, patterns of drug abuse, and correlations with drug abuse.

10. Describe the difference between chronic alcoholism and alcohol abuse.

11. Discuss the three major categories of causes of delinquency:
    a. psychological causes -

    b. sociological causes -

    c. biological causes -

12. Discuss the typologies of adolescent runaways.

KEY TERMS

In your words, provide a definition for each of the following terms:

1. Emotion _____

2. Phobia _____

3. Generalized anxiety disorder _____

4. Self-concept _____

5. Self-esteem _____

6. Proprium _____

7. Physical addiction _____

8. Psychological dependency _____

9. Chronic alcoholism _____

10. Alcohol abuse _____

11. Parricide _____

12. Matricide _____

13. Patricide _____

APPLICATIONS

For each of the following, fill in the blank with one of the terms listed above.

1. A person who is chemically dependent on a drug can be said to have a _____.

2. Whenever Jonathan is asked to answer a question in class, even if he knows the right answer, he becomes so nervous that he can't speak and he begins to sweat and tremble. He may have a _____.

3. A person's opinions of herself, whether good or bad, make up her _____.

4. Cynthia drinks excessively and is chemically dependent on alcohol. She can be said to have _____.

5. Andy has an excessive fear of snakes, to the point where he cannot even look at a picture of a snake without trembling. Andy may have a _____.

6. Jane is very unhappy with herself; she doesn't think she's as smart or as pretty or as socially skilled as some of her friends. Jane has very low _____.

7. A person who is not chemically dependent on a drug, but nonetheless, still craves and needs it, has a _____.

8. _____ is when people drink alcohol to the point where they cannot function in their usual manner and it interferes with aspects of life.

9. Joan exploded in a fit of rage and she shot and killed her father. She has committed _____.

SELF-TEST MULTIPLE CHOICE QUESTIONS

Circle the best answer for each question.

1. Emotions are important because
   a. they affect physical well-being and health.
   b. they affect how people act with others.
   c. they can be sources of pleasure.
   d. all of the above

2. Which of the following emotions would be considered an inhibitory emotion?
   a. affection
   b. fear
   c. anger
   d. jealousy

3. As a child, Nicole tended to be emotionally unresponsive and distant toward people she did not know well. As an adolescent in similar situations,
   a. she is likely to continue this same emotional pattern.
   b. it is difficult to predict how she would react since there is little continuity between childhood and adolescent patterns of emotional responses.
   c. she is likely to react in a hostile manner, since adolescents become more hostile after going through pubertal changes.
   d. she is most likely to now respond in a friendly, warm manner.

4. Which of the following fears is a child likely to outgrow during adolescence?
   a. fear of failure
   b. fear of rejection
   c. fear of public speaking
   d. fear of thunderstorms

5. Dan insists on living on the ground floor rather than any higher floors because he is afraid of heights and would not be able to look out his windows. He has which type of phobia?
   a. agoraphobia
   b. acrophobia
   c. hydrophobia
   d. hematophobia

6. Worry and anxiety can be distinguished from fear because
   a. they are hostile states rather than inhibitory states.
   b. they are pathological states rather than normal states.
   c. they can arise from imaginary scenarios as well as real situations.
   d. they are directed only at specific people.

7. In a study done at the University of Maine by Rice, what was the number one concern of the young women who participated in the study?
   a. grades
   b. getting along with their parents
   c. getting a good job
   d. meaningful social relationships

8. Incidents that make adolescents angry
   a. most likely have to do with conflicts with their parents.
   b. most likely are social in nature.
   c. are more likely to be about objects than people.
   d. are the same sort of incidents that arouse fear in them.

9. Hatred can be a more serious emotion than anger because
   a. it can persist over a longer period of time.
   b. it is difficult to suppress.
   c. it may be expressed in a hostile, aggressive manner.
   d. all of the above

10. Self-concepts are
    a. uni-dimensional, involving only the role in which the individual is currently engaged.
    b. accurate portrayals of the real self.
    c. constantly being formulated over the lifespan.
    d. none of the above

11. A fight with her boyfriend made Mary feel badly about herself for a little while. In this case, which dimension of the self, as suggested by Strang, was affected?
    a. the general self-concept
    b. a temporary self-concept
    c. the conceptualized ideal self
    d. the pragmatic self

12. Nicole gets along well with others and is very accepting of others. She is most likely to have
    a. high self-esteem.
    b. low self-esteem.
    c. unusually high standards of what her ideal self should be.
    d. unusually low standards of what her ideal self should be.

13. An individual with low self-esteem may be more at risk than an individual with high self-esteem
    a. to experience psychosomatic symptoms.
    b. to become a drug abuser.
    c. to become an unwed mother.
    d. all of the above

14. Which of the following is a true statement about the relation between achievement and self-esteem?
    a. High self-esteem can contribute to success in school.
    b. Success in school can contribute to an individual having a positive self-concept.
    c. The relation between school achievement and the self-concept begins early on in school.
    d. all of the above

15. A person's self-concept is most benefitted by
    a. overidentifying with the mother.
    b. overidentifying with the father.
    c. minimal parental identification.
    d. identifying closely with a parent, but not overidentifying with either parent.

16. Which of the following parents are most likely to have a child with high self-esteem?
    a. parents who are very restrictive and critical of their child
    b. parents who pressure their child to achieve in school
    c. parents who are firm and emotionally warm
    d. parents who are permissive

17. According to a research study on remarriage, which of the following children is most likely to have higher self-esteem?
    a. a child living in a reconstituted family who gets along with his stepparent
    b. a child whose parents have divorced and not remarried
    c. a child living in a reconstituted family who does not get along with his stepparent
    d. all of these children will have low self-esteem

18. Children from minority groups are most likely to have low self-esteem if
    a. they have maintained close friendships.
    b. they have experienced white prejudices.
    c. their parents have high self-esteem.
    d. none of the above

19. Which of the following children is most likely to have low self-esteem?
    a. an 8-year-old in elementary school
    b. a 12-year-old in elementary school
    c. a 12-year-old in junior high school
    d. a 16-year-old in high school

20. Often when Beth studies for exams, she takes stimulants to keep her awake so that she can study more. Her pattern of drug use would be classified as
    a. social-recreational
    b. experimental
    c. circumstantial-situational
    d. intensified

21. In comparison to adolescents who are nonabusers of drugs, drug abusers
    a. are more likely to live with both parents.
    b. are more likely to be close to their parents.
    c. are more likely to have parents who disagree about child-rearing practices.
    d. have parents who are more likely to praise and encourage them.

22. Mark was exposed to a lot of violence on television, and it influenced him to become more violent and to commit an illegal aggressive act. This cause of juvenile delinquency would be classified as
    a. psychological
    b. sociological
    c. socioeconomic
    d. biological

23. Which of the following may be a biological cause of juvenile delinquency?
    a. drug abuse
    b. slow-responding autonomic nervous systems
    c. quick-responding autonomic nervous systems
    d. overly developed frontal lobes of the brain

24. Paula ran away from home because her parents set a very early curfew. She went out with her friends to a bar, and then slept over a friend's house without telling anyone else of her whereabouts. She would be classified as which type of runaway?
    a. social pleasure seeker
    b. runaway manipulator
    c. runaway retreatist
    d. endangered runaway

25. Steven ran away from home because his parents set a very early curfew. He went to a section of town where it was very dangerous, but he refused to return home until his parents changed the curfew. He would be classified as a
    a. social pleasure seeker
    b. runaway manipulator
    c. runaway retreatist
    d. endangered runaway

ANSWER KEY

APPLICATIONS

1. physical addiction
2. generalized anxiety disorder
3. self-concept
4. chronic alcoholism
5. phobia
6. self-esteem
7. psychological dependency
8. Alcohol abuse
9. patricide

MULTIPLE CHOICE

1. d (p. 396)	9. d (p. 403)	17. a (p. 408)
2. b (p. 397)	10. c (p. 404)	18. b (p. 410)
3. a (p. 397)	11. b (p. 404)	19. c (p. 410)
4. d (p. 398)	12. a (p. 405)	20. c (p. 412)
5. b (p. 398)	13. d (p. 406)	21. c (p. 413)
6. c (p. 399)	14. d (p. 406)	22. b (p. 416)
7. a (p. 400)	15. d (p. 407)	23. b (p. 417)
8. b (p. 402)	16. c (p. 408)	24. a (p. 419)
		25. b (p. 419)

Chapter 14

SOCIAL DEVELOPMENT

CHAPTER OUTLINE

I. Adolescents in Their Families - The family is the chief socializing influence on adolescents.

    A. What Adolescents Expect of Parents - Some of the things that adolescents have reported expecting of their parents are: reasonable freedom and privileges, faith and trust, approval, willingness to communicate, concern and support, guidance, a happy home, and a good example.

    B. Paternal Involvement - Fathers are more involved with their sons than with their daughters.

    C. Parent-Adolescent Disagreements
       1. When disagreements occur, it usually is in one or more of the following areas: moral-ethical behavior, relationships with family members, academics, fulfilling responsibilities, social activities, and work outside the home.
       2. A number of factors relate to the focus and extent of conflict with parents such as the type of discipline that parents use, the SES of the family, the number of children in the family, the stage of development, and gender.

    D. Parental Functions and Support - Parental functions include: meeting basic needs, protection, guidance, and advocacy. Social supports include: neighborhood and work influences and broader influences such as cultural values.

    E. Relationships with Siblings - Siblings often provide friendship and companionship if they are less than 6 years apart, but there can also be sibling rivalry, particularly in early adolescence.

II. Social Relationships

    A. Friendships of Young Adolescents - Adolescents want emotional independence from their parents and more emotional fulfillment from friends, which can sometimes be disappointing. Females tend to value closeness in relationships, whereas males tend to have difficulty with intimacy.

B. Heterosocial Development - Heterosociality means forming friendships with people of both sexes.

C. Group Involvement
1. Finding acceptance in social groups becomes a powerful motivation in the lives of adolescents.
2. Skills such as conversational ability, ability to empathize with others, and poise are the most important factors in social acceptance, although appearance and achievement play roles as well.
3. Although delinquent or antisocial behavior may be unacceptable in society as a whole, it may be received more favorably in some gangs or other groups.

D. Dating
1. Some of the purposes of dating are: to have fun, for friendships and affection, to maintain status, as a means of personal and social growth, to become more sex oriented, to find intimacy, and later, to find mates.
2. The most frequently mentioned dating concerns for men are: communication, where to go and what to do on dates, shyness, money and honesty/openness.
3. The most frequently mentioned dating concerns for women are: unwanted pressure to engage in sexual behavior, where to go and what to do on dates, communication, sexual misunderstandings, and money.

III. Premarital Sexual Behavior

A. Sexual Interests - Gradually young adolescents become interested not only in their own sexual development, but also in the opposite sex.

B. Masturbation - One of the common practices of adolescents is masturbation which is a normal part of growing up and exploring one's sexuality.

C. Premarital Sexual Intercourse - Since the early 1970's, there has been a trend toward an increase in premarital sexual intercourse. This increase may be an expression of cultural norms.

D. Use of Contraceptives - Large numbers of adolescents do not use any form of birth control and thus are not protected against unwanted pregnancy.

E. Potential Problems
1. The net increase in premarital sexual intercourse accompanied by a lack of efficient use of contraceptives has resulted in an increase in the incidence of out-of-wedlock pregnancies.
2. Adolescents who are sexually active may be exposed to sexually transmitted

diseases, such as venereal diseases and AIDS.
   3. In surveys taken of undergraduates, both males and females report a lot of unwanted sexual activity.

F. Adolescent Marriage
   1. The median age at first marriage has been on the rise since the 1950's.
   2. Numerous studies indicate that the younger people are when married, the greater the chance of unhappy marriage and of divorce.
   3. The most influential motivations for adolescents to marry are: overly-romantic views of marriage, social pressure, acceleration of adult sophistication, sexual stimulation and unwed pregnancy, and escape of personal problems.
   4. Many of the adjustments young couples must make are more difficult because of the immaturity of the couple.

IV. The Development of Moral Judgement

A. Lawrence Kohlberg - Kohlberg identified three major levels of development of moral judgement, each level with two sublevels of motivation.
   1. At Level I, the premoral level, children respond to the definitions of good and bad provided by parental authority figures, and decisions are made on the basis of self-interest. Type 1 obeys rules to avoid punishment. Type 2 conforms to obtain rewards.
   2. At Level II, the level of morality of conventional role conformity, moral judgement is based on a desire to justify, support, and maintain the existing social structure. The child of type 3 conforms to avoid disapproval of others. Type 4 conforms because of a desire to maintain law and order.
   3. At Level III, the level of morality of self-accepted moral principles, individuals accept democratically recognized principles of universal truths because they believe in these principles. Type 5 defines moral thinking in terms of general principles such as mutual obligations, contractual agreements, equity, human dignity, and individual rights. Type 6 is motivated to uphold universal principles of justice that are valid beyond existing laws.
   4. Kohlberg believes that the stages are universal and moved through in succession.
   5. Research has revealed that not everyone makes it to Level III.

B. Carol Gilligan
   1. Gilligan suggested that females approach moral issues from a different perspective than males: women rely on an interpersonal network of care orientation, and men rely more heavily on a justice orientation.
   2. Gilligan proposed three levels of moral reasoning of women: At Level I, women are concerned with survival and self-interest. At Level II, the need to please others takes precedence over self-interest. At Level III, which many

people never attain, the concern is for the consequences for all, including themselves.

    C. A Theory of Reasoned Action - Ajzen and Fishbein suggest that a person's attitudes and beliefs about a specific behavior, as well as social norms, affect behavioral intentions.

V. Work - Another important part of socialization is to learn to work and hold responsible positions, although there is some concern that some adolescents may work too much.

LEARNING OBJECTIVES/ STUDY QUESTIONS

After reading this chapter, you should be able to:

1. Discuss what adolescents expect of their parents.

2. Describe the major sources of conflict between adolescents and their parents.

3. Describe the variables that affect parent-adolescent conflict.

4. Describe the four parental functions. Discuss parents' system of social supports.

5. Discuss adolescent-sibling relationships.

6. Discuss the importance of:
   a. friendships

   b. heterosocial development

   c. group involvement

7. Describe the present trends in the incidence of premarital sexual intercourse and in the use or nonuse of contraceptives.

8. Discuss the issue of sexually transmitted diseases among adolescents.

9. Discuss the problem of unwanted sexual activity among adolescents.

10. Discuss adolescent marriages in terms of frequency, success rates, motivations, and problems of immaturity.

11. Describe Kohlberg's theory of the development of moral judgement and give a critique of the theory.

12. Discuss Gilligan's theory of the moral development of women.

13. Discuss some of the issues involved in adolescent employment while still in school.

KEY TERMS

In your own words, provide a definition for each of the following terms:

1. Imaging_____

2. Masturbation_____

3. Coitus_____

4. Premoral level_____

5. Morality of conventional role conformity_____

6. Morality of self-accepted moral principles_____

APPLICATIONS

For each of the following, fill in the blank with one of the terms listed above.

1. Monica decided to help her friend cheat on a homework assignment because she believed that the friend would help her out at a later time. According to Kohlberg, Monica is at the _____ of moral judgement.

2. Ken is usually loud and tends to dominate conversations. However, when he went on a date with Leslie he made an effort to speak more softly and listen to what she had to say. Ken was _____.

3. John actively fights for equal rights and against human suffering in all parts of the world. He is probably at Kohlberg's level of _____.

4. Another word for sexual intercourse is _____.

5. Ron chose to help out with a charitable event because he didn't want others to think that he was a selfish person. He is probably at Kohlberg's level of
_____.

SELF-TEST MULTIPLE CHOICE QUESTIONS

Circle the best answer for each question.

1. Which of the following best describes the general pattern that has been found in the development of assertiveness during adolescence?
   a. Assertiveness is at its peak at puberty as young adolescents try to establish independence from their parents.
   b. Assertiveness is at its peak in late adolescence as they begin to prepare for adulthood.
   c. Assertiveness increases from early puberty to middle adolescence, after which it tapers off.
   d. Assertiveness tends to be at a constant high level through all of adolescence.

2. Which of the following has generally not been reported by adolescents as being desirable?
   a. parents who give them excessive freedom
   b. parents who show concern and support
   c. parents who are willing to communicate
   d. none of the above

3. Which of the following parents would likely be the most successful at providing guidance and discipline?
   a. a parent who uses physical punishment
   b. a parent who takes away the adolescent's privilege to use the car
   c. a parent who threatens the adolescent with some sort of punishment
   d. a parent who talks to the adolescent and explains why what s/he did was wrong and why s/he should be punished.

4. Which of the following areas of disagreement is probably most common for adolescents and their parents to experience?
   a. conflict over relationships with other family members
   b. conflict over social activities
   c. conflict over work outside the home
   d. conflict over responsibilities at home

5. In terms of their adolescent children, parents of low socioeconomic status tend to be more concerned than middle-class parents with
   a. obedience.
   b. achievement.
   c. initiative.
   d. grades.

6. _____ report more conflicts within their families than do _____.
   a. boys; girls
   b. girls; boys
   c. They report about the same number of conflicts.

7. Which of the following pairs of siblings are most likely to be friends?
   a. siblings who are less than 6 years apart
   b. siblings who are more than 6 years apart
   c. a sibling pair in which the older sibling has more responsibility for caring for the younger siblings than do the parents
   d. siblings who are close in age, but one of them is more favored by their parents

8. Which of the following scenarios is an example of a heterosocial relationship?
   a. Brooke becomes best friends with Amy and they do everything together.
   b. Ted and Bryan are close friends who spend a lot of time talking about girls that they like.
   c. Karl and Michelle enjoy going to baseball games together and often eat lunch together with some of their other friends.
   d. Debbie likes to spend time with Marianne, Diane, Laurie, and Judy, but she doesn't consider any one of them her best friend.

9. An adolescent's need to be involved in a social group is primarily motivated by
    a. a desire to be accepted by others.
    b. a desire to have as many friends as possible.
    c. the need to avoid their parents.
    d. wanting to avoid dating.

10. Which of the following adolescents is most likely to be considered popular?
    a. an attractive one
    b. one who has good social skills
    c. one who does well in school
    d. one who often gets picked on by bullies

11. Which of the following is considered to be an important aspect of dating?
    a. having fun
    b. maintaining status
    c. finding intimacy and affection
    d. all of the above

12. In a study which looked at undergraduates' concerns over dating, which of the following problems was frequently listed by men but not by women?
    a. money
    b. where to go and what to do on a date
    c. honesty, openness
    d. sexual misunderstandings

13. When an adolescent first begins to sexually mature, interest in sexual matters usually initially focuses on
    a. individuals of the opposite sex
    b. individuals of the same sex
    c. one's own bodily changes
    d. the parent of the opposite sex

14. Masturbation by adolescents may
    a. provide sexual release.
    b. provide a way to learn how to respond sexually.
    c. stimulate guilt if the adolescent believes it is harmful or wrong.
    d. all of the above

15. The most popular method of birth control among adolescents who are sexually active and use birth control is
    a. the diaphragm
    b. withdrawal
    c. the pill
    d. condoms

16. Which of the following sexually transmitted diseases is the most common among adolescents?
    a. gonorrhea
    b. genital herpes
    c. syphilis
    d. chlamydial infections

17. Few cases of active AIDS are reported for adolescents because
    a. adolescents are less at risk for AIDS than are adults.
    b. the incubation period for AIDS may be from a few years up to ten years.
    c. adolescents are much less likely to report it.
    d. adolescents engage in less sexual behavior and have fewer problems with drugs than do adults.

18. Based on divorce statistics, men who wait to marry until they are at least _____ years, and women who wait until they are at least _____ years have the best chances of not getting divorced.
    a. 21; 20
    b. 25; 22
    c. 27; 25
    d. 35; 32

19. Which of the following is commonly reported by couples who married young as being a problem in their marriage?
    a. lack of faithfulness
    b. inability to communicate
    c. financial problems
    d. all of the above

20. A child who decides not to steal a piece of candy from a store only because he is afraid that he would be caught would be at which of Kohlberg's developmental levels?
    a. Level I, type 1
    b. Level I, type 2
    c. Level II, type 3
    d. Level II, type 4

21. Margaret is rushing her husband Phil to the hospital after an accident at home. Their daughter Ann tells her to slow down and stop speeding because it is against the law to speed, even though her father needs medical attention. Ann's moral reasoning would put her at which of Kohlberg's levels?
    a. Level I, type 1
    b. Level I, type 2
    c. Level II, type 3
    d. Level II, type 4

22. According to Kohlberg,
    a. an individual cannot get to Level III without first going through Levels I and II.
    b. individuals all over the world will go through the same sequence of moral reasoning.
    c. some individuals may not reach Level III.
    d. all of the above

23. Gilligan's major criticism of Kohlberg's studies of moral judgements is that
    a. he looked at adults rather than at adolescents.
    b. he initially studied only males.
    c. he only studied Americans.
    d. all of the above

24. According to Gilligan, women rely more on _____ than do males in making moral judgements.
    a. an orientation towards care
    b. an orientation towards justice
    c. an orientation towards themselves
    d. a consideration of law and order

25. Sharon decided to volunteer to organize a fund-raising event that no one else had the time for, even though she herself was extremely busy. She told herself that it was more important to help out than to take time for herself. She would be at which level of moral reasoning according to Gilligan?
    a. Level I
    b. Level II
    c. Level III
    d. Level IV

ANSWER KEY

APPLICATIONS

1. premoral level
2. imaging
3. morality of self-accepted moral principles
4. coitus
5. morality of conventional role conformity

MULTIPLE CHOICE

1. c (p. 426)	9. a (p. 434)	17. b (p. 442)
2. a (p. 426)	10. b (p. 435)	18. c (p. 446)
3. d (p. 428)	11. d (p. 436)	19. d (p. 446)
4. b (p. 429)	12. c (p. 437)	20. a (p. 448)
5. a (p. 429)	13. c (p. 438)	21. d (p. 448)
6. a (p. 430)	14. d (p. 439)	22. d (p. 449)
7. a (p. 432)	15. c (p. 441)	23. b (p. 449)
8. c (p. 433)	16. d (p. 442)	24. a (p. 449)
		25. b (p. 450)

Chapter 15

PERSPECTIVES ON ADULT DEVELOPMENT

CHAPTER OUTLINE

I. Demographics

    A. Age Periods - In this book, adulthood has been divided into 3 age groups: early adulthood (the 20s and 30s), middle adulthood (the 40s and 50s), and late adulthood (age 60 and over).

    B. Population Trends - The median age of the U.S. population continues to increase as life expectancy increases, and mortality and fertility rates decline.

    C. Implications
        1. The depression and war-period babies succeeded as adults, primarily because of their small numbers and superior opportunities, in contrast to the baby boomers, for whom there is much more competition for jobs.
        2. The dependency ratio, the number of dependents for each person in the labor force, will continue to increase.
        3. Child dependency and female dependency have been decreasing, while male dependency increases.

    D. Some Positive Developments and Challenges
        1. The divorce rate has leveled off and is decreasing slightly.
        2. The aging of the population has caused a shift from a youth-oriented culture to an adult-centered society.
        3. One challenge that remains is to alter our age-appropriate norms of behavior and make the necessary psychological adjustments to living a longer, more active life.
        4. Society needs a new orientation to the concept of who is old.

II. Meaning of Adulthood

    A. Social Dimensions - The primary meaning of adulthood is social, in that one is perceived as an adult by others.

B. Biological Dimensions - An adult can be defined as one who has attained full size and strength, but this is not adequate enough because many who would still be considered adolescents have achieved their full size.

C. Emotional Dimensions - Being an adult also includes emotional maturity, emotional stability, and functioning autonomously.

D. Legal Dimensions - Laws attempt to differentiate who should and should not be accorded adult rights and responsibilities.

III. Transition to Adulthood

A. Difficulties - Becoming an adult is a complicated process, especially in a pluralistic and highly industrialized society.

B. Passages and Rites - In our culture, numerous rites of passage take place before adulthood can be reached.

C. Socialization
   1. Part of socialization is anticipatory: preparing for certain tasks.
   2. At other times, resocialization is required.

IV. Developmental Tasks

A. The Twenties and Thirties
   1. Detaching oneself from parents, which may involve establishing a separate residence and achieving emotional autonomy, is an important step.
   2. This helps individuals form their personal identities.
   3. A major task of becoming a mature adult is developing the capacity to tolerate tensions and frustration.
   4. Young adults are also involved in making career and education decisions and becoming economically independent.
   5. According to Erikson, the chief psychosocial task is the achievement of intimacy.
   6. Young adults must learn to manage and maintain their own residences.
   7. Many young adults become parents and begin raising families.

B. Middle Age
   1. Physical changes require psychological adjustments and adjustments in lifestyle and health habits.
   2. Usually, middle age is the most fruitful period of professional and creative work.
   3. Participation in community life is essential for society's progress.
   4. Part of the developmental task at this point is to let go of the responsibility

for, as well as the control of, the children.
5. Couples whose children have left home face the task of working out problems and becoming close again.
6. Responsibility for providing assistance to aging parents may increase.
7. Once children are independent, crossing of adult sex roles becomes more apparent.
8. There is an increasing need for couple-centered activities and an increased interest in having fun, pursuing interests, and finding new meaning in life.
9. One research study suggests that for women, personality changes in normative ways during middle adulthood. For example, women become less dependent and self-critical and more confident and decisive.

C. Late Adulthood
1. The task of staying physically healthy becomes more difficult.
2. Adequate income may be a problem.
3. Keeping one's own residence may be important.
4. Older adults may feel a loss of status when they retire. It is helpful if they set new goals and are able to maintain a comfortable lifestyle.
5. Their challenge is to find meaningful relationships with others and to adjust to new family roles.
6. According to Erikson, the development of ego integrity is the chief psychosocial task of the final stage of life.

V. Theories of Adult Development Over the Life Span

A. Gould - Phases of Life: Gould outlined 7 phases of life, beginning at age 16, of the changes and adjustments required as people age.

B. Levinson - Seasons of Life
1. Based on a study with 40 men, Levinson proposed a model of adult development that included periods of relative stability interspersed with periods of transition at ages 22, 40, and 60.
2. He contends that women have life stages similar to those experienced by men, but that women face more serious life problems than men during the transition periods.

C. Vaillant - Adaptation to Life
1. Vaillant conducted a longitudinal study of some of the "best and the brightest" white males.
2. He found that the quality of childhood environments was related to outcomes, but it was not the sole determinant of success, that negative traits during adolescence did not predict adult outcomes, and that achieving intimacy during young adulthood was important for later marital success.

D. Comparison and Critique of Studies
      1. The age divisions in the three studies are purely arbitrary.
      2. All three researchers described a period of transition between adolescence and early adulthood, and a mid-life crisis around age 40.
      3. These studies involved only middle- and upper-middle-class segments of the population, and only Gould's study included women, therefore these results may not be generalizable.

VI. Causes of Change and Transition

   A. Normative-Crisis Model - This model describes human development in terms of a definite sequence of age-related biological, social, and emotional changes.

   B. Timing-of-Events Model - This model suggests that development is not the result of a set plan or schedule of crises but is a result of the time in people's lives when important events take place.

LEARNING OBJECTIVES/ STUDY QUESTIONS

After reading this chapter, you should be able to:

1. Discuss the demographics of adulthood and the implications for our society.

2. Describe the positive changes taking place in our society as a result of the aging of the population.

3. Summarize the social, biological, emotional, and legal dimensions of adulthood.

4. Discuss the difficulty of making the transition to adulthood in our society as compared with the transition in primitive cultures.

5. Summarize the developmental tasks of early, middle, and late adulthood.
   a. Early adulthood -

   b. Middle adulthood -

   c. Late adulthood -

6. Summarize the developmental theories of Gould, Levinson, and Vaillant.
   a. Gould -

   b. Levinson -

   c. Vaillant -

7. Compare the theories of Gould, Levinson, and Vaillant and critique them.

8. Discuss the normative-crises model describing change and transition in adult life.

9. Discuss the timing-of-events model that describes the changes and transitions that take place in adult life.

KEY TERMS

In your own words, provide a definition for the following terms:

1. Baby boomers_____

2. Dependency ratio_____

3. Normative-crisis model_____

4. Biological time clock_____

5. Social clock_____

6. Timing-of-events model_____

7. Normative influences_____

8. Nonnormative or idiosyncratic influences_____
_____

APPLICATIONS

For each of the following, fill in the blank with one of the terms listed above.

1. Both puberty and menopause are determined by a _____.

2. The model that suggests that important events in people's lives, such as marriage or divorce, are what stimulates development is the _____.

193

3. An individual born in 1952 would be considered a _____.

4. Denise is graduating from college at the age of 22 and she knows she must look for a job. She has been planning for it for the past few years. This life event would be considered a _____ influence.

5. In our society, the majority of individuals who attend college begin so in their late teens or early twenties, thus following a _____ clock.

6. An individual who believes that adult development can be described in terms of a sequence of age-related changes subscribes to a _____.

7. Eric was unexpectedly fired from his job at the age of 45, thus throwing his life into confusion. This would be considered a _____ influence.

8. If one wants a way to refer to how many people there are in the work force in comparison to how many people are being supported, one could use the _____.

SELF-TEST MULTIPLE CHOICE QUESTIONS

Circle the best answer for each question.

1. The median age of U.S. population
    a. has increased over the past 20 years, but is now beginning to decrease.
    b. has decreased over the past 20 years, but is now beginning to increase.
    c. continues to decrease.
    d. continues to increase.

2. Baby boomers have had to compete more for jobs than did the previous generation because
    a. more jobs have become available due to advanced technology.
    b. more jobs became available after World War II.
    c. there are so many of them competing for jobs.
    d. of the degree of stability in the economy.

3. A dependency ratio of 1.4 to 1 means that
    a. there are on average 1.4 dependents for every person working.
    b. there are on average 1.4 people working for every person not working.
    c. for every married couple, there are on average 1.4 children.
    d. for every adult, there are on average 1.4 children.

4. The divorce rate
   a. is continuing to increase.
   b. is continuing to decrease.
   c. has leveled off and is starting to decrease slightly.
   d. has leveled off and is starting to increase slightly.

5. In our society, which of the following could be considered a rite of passage into adulthood?
   a. graduating from high school
   b. taking a driver's test
   c. voting for the first time after turning 18
   d. all of the above

6. Jane is 28 years old when she moves out of her parents' house and gets an apartment of her own. She is
   a. involved in a developmental crisis.
   b. engaged in an appropriate developmental task.
   c. developmentally delayed.
   d. none of the above

7. In early adulthood, more important than physical separation from parents is
   a. emotional autonomy.
   b. financial independence.
   c. material independence.
   d. dependency on peers.

8. According to Erikson, the chief psychosocial task of early adulthood is
   a. the formation of personal identity.
   b. finding autonomy.
   c. the achievement of intimacy.
   d. finding a career.

9. Marital satisfaction usually is at its lowest point
   a. right after the first child is born.
   b. right after a second child is born.
   c. when the children reach school age.
   d. when the children reach adolescence.

10. One important developmental task during middle adulthood is
    a. letting go of the responsibility and control of the children.
    b. achieving autonomy.
    c. establishing a residence.
    d. all of the above

11. Adults usually become more responsible for their _____ and less responsible for their _____ during the middle-age years.
    a. children; parents
    b. parents; children
    c. spouses; parents
    d. children; spouses

12. Often in middle adulthood, adult sex roles become less pronounced because
    a. women become more dependent as they get older.
    b. as the children get older, many women return to the work force.
    c. the men are not as pressured at work.
    d. all of the above

13. Which of the following is often a concern in late adulthood?
    a. staying physically fit
    b. having adequate income
    c. being able to keep one's own home
    d. all of the above

14. Which of the following people is most likely to be satisfied with retirement?
    a. a person who decided on the spur of the moment to retire
    b. a person who has to retire because of mandatory retirement rules
    c. a person who retires because he hates his job
    d. a person who has planned for his retirement

15. One of the potential problems that older adults must face when they retire is
    a. feeling a loss of status and prestige.
    b. feeling as if they have lost their main identity.
    c. financial problems.
    d. all of the above

16. According to Erikson, the chief psychosocial task of late adulthood is
    a. the development of intimacy.
    b. acceptance of retirement.
    c. the continuation of kinship relations.
    d. the development of ego integrity.

17. According to Gould, adults reach a settling-down stage between the ages of
    a. 22 to 29.
    b. 29 to 35.
    c. 35 to 43.
    d. 43 to 50.

18. According to Levinson, which of the following age periods is transitional?
    a. ages 17 to 22
    b. ages 28 to 33
    c. ages 40 to 45
    d. all of the above

19. Based on Levinson's model of adult development, a person who is upset because he is intensely reexamining his life, and who questions every aspect of his life is probably between the ages of
    a. 33 to 40.
    b. 40 to 45.
    c. 45 to 50.
    d. 50 to 55.

20. According to Levinson,
    a. women face less serious life problems during transition stages.
    b. women are more likely to be satisfied with the amount of intimacy in their lives.
    c. women are less likely to be satisfied when they begin to question the decisions they have made in their lives.
    d. none of the above

21. Vaillant found that
    a. mental illness during childhood predicted adult emotional illness.
    b. childhood environment was the sole determinant of adult success.
    c. negative traits during adolescence predicted the Worst Outcomes as adults.
    d. men with unhappy childhoods were more likely to become mentally ill.

22. Vaillant's study also found that the achievement of _____ is very important during young adulthood.
    a. a successful career
    b. intimacy
    c. autonomy from parents
    d. trust

23. A model that describes human development in terms of a sequence of age-related transitions and stages is a
    a. normative-crisis model.
    b. timing-of-events model.
    c. transitional stage model.
    d. stability model.

24. A _____ is dependent on the specific culture, whereas a(n) _____ is the same all over the world.
   a. social clock; biological clock
   b. maturational clock; social clock
   c. biological clock; emotional clock
   d. social clock; emotional clock

25. According to a timing-of-events model, which of the following is most likely to lead to a turning point in one's life?
   a. becoming 40 years old
   b. preparing for marriage
   c. planning to change one's job
   d. an unexpected job layoff

ANSWER KEY

APPLICATIONS

1. biological time clock
2. timing-of-events model
3. baby boomer
4. normative
5. social
6. normative-crisis model
7. nonnormative or idiosyncratic
8. dependency ratio

MULTIPLE CHOICE

1. d (p. 462)	9. c (p. 472)	17. d (p. 477)
2. c (p. 463)	10. a (p. 473)	18. d (p. 479)
3. a (p. 465)	11. b (p. 473)	19. b (p. 479)
4. c (p. 466)	12. b (p. 473)	20. c (p. 479)
5. d (p. 470)	13. d (p. 475)	21. d (p. 480)
6. b (p. 471)	14. d (p. 475)	22. b (p. 480)
7. a (p. 471)	15. d (p. 475)	23. a (p. 484)
8. c (p. 472)	16. d (p. 476)	24. a (p. 484-85)
		25. d (p. 486)

Chapter 16

PHYSICAL DEVELOPMENT

CHAPTER OUTLINE

I. Physical Attractiveness, Abilities, and Fitness

    A. Growth and Aging
       1. Sometime during the mid-20s, the human body is usually at the peak of its physical development.
       2. One characteristic of middle age is a growing awareness of personal mortality accompanying the first physical signs of aging.

    B. Attitudes of Society
       1. Self-consciousness about one's changing physique is accentuated by society's attitudes, particularly for women.
       2. The elderly are being portrayed on television in a more positive way; however, ageism still exists in our society.

    C. Physical Fitness and Health
       1. Chronological age alone is a poor measure of physical conditioning or aging; the emphasis should be on functional age.
       2. In general, reaction time decreases from childhood to about age 20, remains constant until the mid-20s, and then slowly increases.
       3. Most of the decline in motor ability, coordination, and dexterity occurs after the 30s.
       4. Regular exercise can maintain power and even restore it to muscles that have gone unused.
       5. The maximum work rate one can achieve without fatigue begins to decline at about 35.

    D. Exercise
       1. Exercise is one of the best ways to prevent ill health, maintain body fitness, and delay aging decline.
       2. One of the simplest measures of health and physical fitness is pulse rate, which increases during exercise.

E. Diet and Nutrition
   1. Proper nutrition results in feelings of well-being, high energy levels to carry on daily activities, and maximum resistance to disease and fatigue.
   2. The average adult in the U.S. is 20 pounds overweight.
   3. Adults may require special diets, for example, diets low in saturated fats and cholesterol, sodium, and sugar.
   4. With aging, the accumulation of fat in the abdominal area increases. Exercise can improve fat distribution.

F. Rest and Sleep
   1. Sleeping habits change as one ages, and the rest received may not be adequate.
   2. Insomnia may be caused by underlying biological predispositions, psychological factors, the use of drugs and alcohol, disturbing environment and bad habits, and negative conditioning.

G. Marital Happiness and Mismatched Sleep/Wake Cycles - Researchers have found that couples whose sleep/wake patterns are out of sync are more likely to have troubled marriages than couples whose patterns are similar.

H. Drug Abuse
   1. The greatest abusers of drugs are young adults between the ages 18 and 25.
   2. Studies indicate widespread drug abuse among the elderly who report taking them for medical reasons. All kinds of drugs are potentially more harmful for older persons than for younger persons.

II. Some Bodily Systems and their Functioning

   A. Nervous Systems
      1. As aging progresses, brain weight declines, but this is due to a decline in size of brain cells, and not because of cell death.
      2. The brain's efficiency depends primarily on the amount of blood and oxygen it receives.
      3. The most widely accepted explanation for the slowing down of bodily functions with age is that nerve impulses are transmitted more slowly as people age.

   B. Cardiovascular System
      1. Changes in the heart and blood vessels cause a decline in the heart's pumping power and stroke volume.
      2. The health of the heart depends partially on the health of the blood vessels that transport the blood.
      3. Heart disease is the number one killer of the elderly.

C. Respiratory System - Lung efficiency is reduced during aging due to a number of factors. Thus, the elderly cannot tolerate exercise as well.

D. Gastrointestinal System - Problems with liver functioning and gall bladder trouble are common in old age.

E. Urinary System - Diseases of the kidney can be serious because poisons collect in the body if the kidneys function improperly.

F. Skeletal-Dental Systems
  1. One of the most noticeable signs of advancing age is a change in stature and posture.
  2. The risk of all injuries increases with age.
  3. Periodontal disease and tooth decay are primary reasons for tooth loss.

G. Reproductive System
  1. Menopause - The cessation of menstruation takes place over a period of years.
  2. Male climacteric - The decline in reproductive function in the male is generally a gradual process.

III. Human Sexuality

A. Sexual Relationships - Although the frequency of intercourse generally declines for married couples, most couples still maintain interest in sexual relations.

B. Sexual Dysfunction
  1. Any malfunction of the human sexual response system is called sexual dysfunction.
  2. Male sexual dysfunctions include: inhibited sexual desire, ejaculatory inhibition, erectile dysfunction, and premature ejaculation.
  3. Female sexual dysfunctions include: general sexual dysfunction, orgasm dysfunction, vaginismus, and painful intercourse.
  4. Physical and emotional factors can cause sexual dysfunctions. Most difficulties can be cleared up with proper help.

IV. The Senses and Perception

A. Visual Acuity - Visual acuity reaches a maximum around age 20 and remains relatively constant to 40, then begins to decline. Research studies suggest that there is a moderate relationship between aging and automobile accidents, but little effect on reading speed.

B. Hearing Acuity - Hearing ability reaches its maximum around age 20.

C. Taste and Smell - The ability to perceive all four taste qualities declines in later life, although the decrement is small. There is also decline in olfactory function.

D. Tactile Sensitivity: Touch, Temperature, and Pain - There is some decrease in tactile acuity with increasing age, but the loss is small.

E. Thermoregulation - Older adults have more difficulty maintaining body temperature.

F. Sense of Balance - A maximum sense of balance is achieved between 40 and 50, followed by decline.

V. Biological Aging

A. Senescence - Senescence is a term used to describe biological aging.

B. Theories of Biological Aging
1. Hereditary theory - The theoretical length of life is hereditary.
2. Cellular aging theory - Aging is programmed by the limited capacity of cells to replace themselves.
3. Wear-and-tear theory - This theory emphasizes that the organism simply wears out.
4. Metabolic waste or clinker theory - Aging is caused by the accumulation of deleterious substances within various cells of the body.
5. Autoimmunity theory - This theory describes the process by which the body's immune system rejects its own tissues through the production of autoimmune antibodies.
6. Homeostatic imbalance theory - This theory emphasizes the gradual inability of the body to maintain vital physiologic balances.
7. Mutation theory - This theory describes what happens when more and more body cells develop mutations.
8. Error theory - This theory is a variation of the mutation theory which includes the cumulative effects of a variety of mistakes that may occur.
9. No single theory adequately explains the complex events that occur in aging.

VI. Prolongevity

A. Possibilities - The term prolongevity describes deliberate efforts to extend the length of life by human action.

B. Implications - The goal of prolongevitists is not simply to increase the maximum life span, but to retard both disease and the aging process.

LEARNING OBJECTIVES/ STUDY QUESTIONS

After reading this chapter, you should be able to:

1. Discuss when the human body is at the peak of physical development.

2. Discuss the self-consciousness about one's changing physique accentuated by society's attitudes that equate attractiveness with youthfulness.

3. Describe the meaning and some examples of ageism in our society.

4. Discuss functional age vs. chronological age.

5. Discuss the importance of exercise in maintaining body fitness and good health.

6. Discuss the importance of nutrition for good health and discuss diets suitable for adults.

7. Discuss the importance of adequate sleep and rest in maximizing healthful functioning and the most important factors that contribute toward insomnia.

8. Discuss drug abuse among adults.

9. Describe age-related changes and common diseases of the following systems:
   a. nervous system -

   b. cardiovascular system -

   c. respiratory system -

   d. gastrointestinal system -

   e. urinary system -

   f. skeletal-dental system -

   g. reproductive system -

10. Discuss the different types of male and female sexual dysfunctions.

11. Describe age-related changes in the following sensory systems:
   a. vision -

   b. hearing -

   c. tactile acuity, temperature, and pain sensitivity -

   d. taste -

12. Discuss the meaning and process of senescence.

13. Discuss the principle theories of biological aging.

14. Discuss longevity and the goals of prolongevitists.

KEY TERMS I

In your own words, provide a definition for each of the following terms:

1. Ageism _____

2. Functional age _____

3. Reaction time _____

4. Motor ability _____

5. Pulse rate _____

6. Nutrient density _____

7. Basal calories _____

8. Activity calories _____

9. Saturated fats _____

10. Cholesterol _____

11. LDL (low-density lipoprotein) _____

12. HDL (high-density lipoprotein) _____

13. Psychotropics _____

14. Arteriosclerosis _____

15. Systolic blood pressure _____

16. Diastolic blood pressure _____

17. Hypertensive heart disease _____

18. Atherosclerosis _____

19. Angina pectoris _____

20. Coronary occlusion with myocardial infarction _____
_____

21. Ischemic heart disease_____

22. Congestive heart failure_____

APPLICATIONS I

For each of the following, fill in the blank with one of the terms listed above.

1. A buildup of fatty deposits between the middle muscular layer and lining layer of arterial tissue is called _____.

2. The most common and serious diseases of the heart in the elderly are _____.

3. The time interval between hearing one's name called and turning the head in the direction of the voice is called the _____.

4. A person who eats a diet high in animal fats is likely to have a high _____ level.

5. The pressure produced when the chambers of the heart dilate and fill with blood is the _____.

6. Television programs that portray older individuals as being irritable and demanding are engaging in _____.

7. The energy that is expended when an individual jogs is referred to as _____.

8. When a person suffers a complete cutoff of blood from a coronary artery, she can be said to have experienced a _____.

9. _____ is a harmful cholesterol; _____ is a beneficial cholesterol.

10. A person who is physically and mentally fit and is very active probably has a lower _____ than chronological age.

11. The hardening of the arteries by a buildup of calcium in the middle muscle layer of arterial tissue is called _____.

12. A candy bar that has few nutrients, but is high in calories, has a low _____.

KEY TERMS II

In your own words, provide a definition for the following terms:

1. Edema _____

2. Cardiac arrhythmias _____

3. Cerebrovascular disease _____

4. Thrombosis _____

5. Embolism _____

6. Hemorrhage _____

7. Stroke _____

8. Pulmonary thrombosis _____

9. Pulmonary embolism _____

10. Vital capacity _____

11. Alveoli _____

12. Tuberculosis _____

13. Bronchial pneumonia _____

14. Pulmonary infections _____

15. Emphysema _____

16. Glycogen _____

17. Pernicious anemia _____

18. Hemoglobin _____

19. Jaundice _____

20. Cirrhosis _____

21. Bile _____

22. Gallstones _____

23. Insulin _____

24. Diabetes mellitis _____

25. Gastritis _____

26. Incontinent _____
_____

APPLICATIONS II

For each of the following, fill in the blank with one of the terms listed above.

1. The air cells of the lungs are called _____.

2. _____ are irregular heartbeats.

3. A person whose skin and whites of the eyes are yellow in color may have _____.

4. _____ is a blockage of blood vessels to the lungs by a blood clot.

5. _____ is an infectious disease of the lungs, while _____ is an infectious disease of the bronchial tubes.

6. Stones that are formed in the gall bladder or bile passages when the bile becomes overconcentrated are called _____.

7. If the pancreas does not produce sufficient insulin, there may be excess sugar in the system and the individual may suffer from _____.

8. A person who has a blood vessel that ruptures, causing severe bleeding, is experiencing a _____.

9. The volume of air inhaled by the lungs with each breath is called the _____.

10. When the body needs sugar, the liver releases _____ into the bloodstream.

11. When the connective tissue of the liver becomes hard, lumpy, and shriveled, the individual has _____.

12. A person who has a _____ may become paralyzed because the blood supply to part of the brain has been cut off and some brain cells have died.

13. The red pigment of the blood is called _____.

14. _____ is an inflammation of the stomach lining.

KEY TERMS III

In your own words, provide a definition for each of the following terms:

1. Osteoarthritis_____

2. Rheumatoid arthritis_____

3. Periodontal disease_____

4. Osteoporosis_____

5. Menopause_____

6. Climacteric_____

7. Sexual dysfunction_____

8. Inhibited sexual desire_____

9. Ejaculatory inhibition_____

10. Erectile dysfunction_____

11. Premature ejaculation_____

12. General sexual dysfunction_____

13. Orgasm dysfunction_____

14. Vaginismus_____

15. Dyspareunia_____

16. Visual acuity _____

17. Presbyopia _____

18. Accommodation _____

19. Adaptation _____

20. Peripheral vision _____

21. Cataracts _____

22. Glaucoma _____

23. Macular diseases _____

24. Presbycusis _____

25. Tactile acuity _____

26. Thermoregulation _____

27. Senescence _____
_____

APPLICATIONS III

For each of the following, fill in the blank with one of the terms listed above.

1. Even though the human body is exposed to many different external temperatures, we maintain a constant body temperature because we are capable of _____.

2. A person who is able to see small details of an intricate map has good _____.

3. James is very frustrated because he cannot maintain an erection long enough to have intercourse. This is an example of _____.

4. A person with _____ may lose her vision over time because the fluid pressure within the eyeball increases.

5. A person whose lenses of the eyes are cloudy and opaque may have _____.

6. Jade is unable to enjoy intercourse because she experiences spasms and contractions of her vaginal muscles. This is called _____.

7. _____ is a condition in which there is calcification and loss of bone mass.

8. A middle-aged woman who experiences hot flashes, numbness or tingling, headaches, joint pain and dizziness may be experiencing _____.

9. _____ refers to the ability of the eye to open and close the pupil depending upon the amount of light.

10. A person who cannot clearly focus on objects that are a short distance from him may have _____.

11. The process of biological aging is called _____.

12. A person who chronically has painful swelling of the small joints in his hands may have _____.

SELF-TEST MULTIPLE CHOICE QUESTIONS

Circle the best answer for each question.

1. The human body is usually at the peak of its physical development sometime during the
    a. teen years.
    b. mid-20s.
    c. mid-30s.
    d. mid-40s.

2. A 50-year-old who acts like a 70-year-old has a higher _____ age than _____ age.
    a. mental; chronological
    b. chronological; mental
    c. functional; chronological
    d. chronological; functional

3. At which age is a person likely to have the fastest reaction times?
    a. 10 years
    b. 16 years
    c. 25 years
    d. 35 years

4. A person who plays basketball recreationally would most likely notice a decline in skills
   a. after turning 40.
   b. after turning 50.
   c. during the 20s.
   d. during the teens.

5. Exercise can
   a. provide more oxygen to the entire body.
   b. tone the muscles.
   c. improve digestion.
   d. all of the above

6. Victor is very physically fit and regularly exercises, while Mickey is not. If they go out running together, what will happen to their pulse rates?
   a. Mickey's will be higher than Victor's.
   b. Victor's will be higher than Mickey's.
   c. They will increase at the same rate because they are engaging in the same exercise.
   d. They will decrease at the same rate because they are engaging in the same exercise.

7. Adult diets should have food of greater nutrient density because
   a. adults require more calories with age.
   b. energy requirements decrease, but adults still need as many nutrients.
   c. nutritional needs decrease, but energy requirements increase.
   d. more calories and more nutrients are required.

8. Researchers who believe that people are biologically predisposed to insomnia suggest that insomniacs
   a. have overly active hypothalamuses.
   b. have overly active hypnagogic systems.
   c. have overly active arousal systems.
   d. have overly active arousal systems, and underactive hypnagogic systems.

9. Which of the following is a true statement about drug use?
   a. Very few older persons take drugs for nonmedical purposes.
   b. All kinds of drugs are potentially more harmful for younger than for older persons.
   c. The greater the number of drugs taken, the less likely that adverse reactions will occur.
   d. all of the above

10. With age, brain weight declines because
    a. cells die and are not replenished.
    b. nerves move closer together.
    c. the size of the brain cells decreases.
    d. larger cells die and are replaced by smaller, more efficient cells.

11. Bodily functions such as walking and writing slow down with age because
    a. nerve impulses are transmitted more slowly across nerve connections.
    b. the chemical transmitters deteriorate over time.
    c. the chemical transmitters change chemical structure over time.
    d. all of the above

12. Because the walls of the aorta become less elastic with age,
    a. systolic blood pressure naturally decreases.
    b. systolic blood pressure naturally increases.
    c. diastolic blood pressure naturally increases.
    d. diastolic blood pressure naturally decreases.

13. Progressive deposits of calcium in the arteries can cause the arteries to harden, resulting in
    a. a decrease in size of the entire heart.
    b. the heart needing to work less hard to pump blood.
    c. elevated blood pressure.
    d. none of the above

14. The number one killer of the elderly is
    a. cancer.
    b. diabetes.
    c. brain tumors.
    d. heart disease.

15. With age, the breathing rate _____, while breathing capacity _____.
    a. increases; decreases
    b. decreases; increases
    c. increases; remains constant
    d. remains constant; decreases

16. The most important organ in digestion is
    a. the stomach.
    b. the gall bladder.
    c. the liver.
    d. the pancreas.

17. A major cause of cirrhosis of the liver is
    a. consumption of alcohol.
    b. cigarette smoking.
    c. air pollution.
    d. all of the above

18. Many older individuals become incontinent after having a stroke which means that they
    a. are paralyzed.
    b. need to have a special diet.
    c. cannot control their bladder.
    d. are mentally impaired.

19. Which of the following statements about menopause is true?
    a. Menopause is a disease which can be treated with proper medication.
    b. All women have gone through menopause by age 48.
    c. About 25% of menopausal women experience symptoms such as hot flashes, headaches, dizziness, joint pain, and bladder difficulties.
    d. Most symptoms of menopause are psychosomatic.

20. Which of the following is the best statement about sexual dysfunctions and orgasms?
    a. Both men and women are likely to have problems reaching climax.
    b. A common problem for women is that they cannot reach climax, but this is not a problem for men.
    c. One of the most common sexual dysfunctions for men is the inability to reach climax, but women rarely have this problem.
    d. A common sexual dysfunction for women is the inability to reach climax; this is less common for men, but does occur.

21. Which of the following types of sexual dysfunctions can be experienced by either men or women?
    a. lack of sexual desire
    b. erectile dysfunction
    c. dyspareunia
    d. none of the above

22. An individual with presbyopia
    a. cannot see clearly up close.
    b. cannot see clearly objects in the distance.
    c. need brighter light than normal to see clearly.
    d. have cloudy vision at any distance.

23. An older individual with presbycusis would have the most difficulty hearing the word
    a. car.
    b. rob.
    c. man.
    d. fat.

24. *Senescence*, in contrast to *senility*,
    a. is a natural occurrence.
    b. is a disease.
    c. is not inevitable.
    d. all of the above

25. According to which theory of aging is aging programmed by the limited capacity of cells to replace themselves?
    a. heredity theory
    b. cellular aging theory
    c. metabolic waste theory
    d. autoimmunity theory

26. According to which theory of aging is aging caused by the accumulation of deleterious substances within various cells of the body?
    a. wear-and-tear theory
    b. homeostatic imbalance theory
    c. mutation theory
    d. metabolic waste theory

27. Which of the following individuals could be described as a prolongevitist?
    a. one who believes that people have the right to choose when they want to die
    b. one who works to extend the average length of life by human actions
    c. one who works to improve the quality of life for older individuals by reducing their pain through medication
    d. one who believes humans should not interfere with the natural process of death

ANSWER KEY

APPLICATIONS I

1. atherosclerosis
2. ischemic heart diseases
3. reaction time
4. cholesterol
5. diastolic blood pressure
6. ageism
7. activity calories
8. coronary occlusion with myocardial infarction
9. LDL; HDL
10. functional age
11. arteriosclerosis
12. nutrient density

APPLICATIONS II

1. alveoli
2. Cardiac arrhythmias
3. jaundice
4. Pulmonary embolism
5. Tuberculosis; bronchial pneumonia
6. gallstones
7. diabetes mellitis
8. hemorrhage
9. vital capacity
10. glycogen
11. cirrhosis
12. stroke
13. hemoglobin
14. Gastritis

APPLICATIONS III

1. thermoregulation
2. visual acuity
3. erectile dysfunction
4. glaucoma
5. cataracts
6. vaginismus
7. Osteoporosis
8. menopause
9. Adaptation
10. presbyopia
11. senescence
12. rheumatoid arthritis

MULTIPLE CHOICE

1. b (p. 492)
2. c (p. 495)
3. c (p. 496)
4. a (p. 497)
5. d (p. 498)
6. a (p. 499)
7. b (p. 501)
8. d (p. 504)
9. a (p. 505)
10. c (p. 507)
11. a (p. 507)
12. b (p. 508)
13. c (p. 508)
14. d (p. 509)
15. d (p. 509)
16. c (p. 509)
17. a (p. 510)
18. c (p. 510)
19. c (p. 512)
20. d (p. 514)
21. a (p. 514)
22. a (p. 515)
23. d (p. 517)
24. a (p. 519)
25. b (p. 520)
26. d (p. 520)
27. b (p. 521)

Chapter 17

COGNITIVE DEVELOPMENT

CHAPTER OUTLINE

I. Cognitive Development

   A. Formal Operational Thinking
      1. Formal operational thinking involves the thought processes of introspection, abstract thinking, logical thinking, and hypothetical reasoning.
      2. Approximately half of the adult population may never attain the full stage of formal thinking, and some adults are better able to use formal thinking in their field of specialization, but not in other fields.
      3. There is some evidence that older adults approach problems at a lower level of abstraction, but this may be because they approach problems differently.

   B. Practical Problem-Solving Abilities - Some research has found a diminution with age in capacity to do abstract problems, but an improvement in ability to solve practical problems that might actually be encountered.

   C. Comprehension
      1. Tests of word familiarity have shown that adults generally show performance improvement through the 50s, after which scores decline. However, educational level was found to be more important than age itself.
      2. The ability to comprehend relativized sentences remains stable until the 60s, after which it declines.
      3. Older adults were able to comprehend the meanings of short prose passages almost as well as younger adults.

   D. Wisdom - One of the advantages of getting older is that people develop pragmatic knowledge we call wisdom.

   E. Problem Finding - Cognitive growth in adulthood is continuous; there is no end point beyond which new structures may appear.

   F. Dialectical Thinking - Some adults are better at an advanced form of thought

218

called dialectical thinking, which involves being able to consider both sides of an issue simultaneously.

G. Schaie's Stages of Adult Cognitive Development - Schaie identified five stages of acquiring knowledge and making increased use of it.

II. Intelligence

A. Scores on the WAIS as a Function of Age - In general, verbal scores tend to hold up with increasing age, whereas performance scores tend to decline after the mid-20s.

B. Cross-Sectional Versus Longitudinal Measurements - Both longitudinal and cross-sectional measurements show that verbal scores remain the most stable and performance scores decline the most.

C. Scores on the PMA as a Function of Age - Intellectual decline was less when measured longitudinally than cross-sectionally. Decline does not always occur, but when it does, it usually takes place after age 50.

D. Fluid and Crystallized Intelligence
  1. Horn and Cattell found that fluid intelligence declined after age 14 with the sharpest decline in early adulthood, while crystallized intelligence showed increases through adulthood.
  2. Schaie argued that the apparent decline in fluid intelligence was due to generational differences.

E. Criticisms
  1. IQ is only one important factor necessary to carry out responsible tasks.
  2. Some test items are more familiar to children or very young adults, and tests may show a cultural and economic bias.
  3. IQ does not measure innate capacity; it is adjusted for the individual's age.

F. Factors Affecting Scores
  1. Adults score better if test items are relevant to their daily lives.
  2. The complexity of the task, the degree of motivation, personality traits, physical factors, emotional factors, and hearing acuity can all affect performance.
  3. The relationship between the test administrator and the test takers can affect performance.
  4. General intellectual level and years of school completed are also related to test performance.

G. Personality, Behavior, and Mental Abilities
1. There is a positive relation between mental health and cognitive ability.
2. Personality indirectly affects intellectual functioning by influencing life cycle changes.
3. Individuals become more cautious with time, which may be related to test performance.
4. An individual with more intelligence and education will exhibit less rigidity.
5. The maintenance of intellectual abilities with advancing age partly depends on what people expect will happen.

H. Socioenvironmental Effects
1. Older adults exposed to intellectually stimulating environments maintain a higher level of cognitive ability with increasing age.
2. People with high education levels and superior socioeconomic status show less decline in cognitive abilities with age, especially for verbal abilities.
3. Mental deterioration is also related to the frequency and intensity of life crises.
4. Abilities most dependent on acculturation are considered the most trainable.

I. Terminal Decline - The theory of terminal decline holds that many human functions not only are primarily related to chronological age, but also show marked decline during a period of a few weeks to a few years before death.

III. Information Processing

A. Memory
1. Memory can be divided into prospective memory (geared toward the future) and retrospective memory (memory for past events).
2. The three basic processes of memory are acquisition, storage, and retrieval.
3. Memory storage consists of sensory memory, short-term or primary memory, and long-term or secondary memory.
4. Age-related differences in memory are not great.
5. Sensory memory abilities of older adults depend partially on the extent to which their sensory receptors are functioning at normal levels.
6. Tactual memory declines faster than visual or auditory memory.
7. Primary memory span does not change with age or it decreases only slightly.
8. Many studies show that older subjects perform less well than younger subjects when secondary memory is involved, but aging does not always bring memory deficits.
9. Older subjects are at their greatest disadvantage when materials to be learned are meaningless or unfamiliar.
10. Studies show that older subjects are poorer than younger ones at facial recognition memory and recalling and recognizing events from the past.
11. One study found that spatial cognitive ability affects older adults' use of their neighborhoods.

12. There are three kinds of long-term memory storage: procedural, semantic, and episodic. Age-related changes are found primarily in episodic memory.
13. Evidence indicates that memory of older adults can be improved with training.
14. Subjects of all ages have been found to remember best sociohistoric events that occurred when they were 15 to 25 years old.

B. Learning
1. Much of what was previously regarded as learning ability deficiency in later life is now seen as a problem in the ability to express learned information.
2. Verbal learning research uses the tasks of paired-associate learning, serial learning tasks, and divided attention tasks.
3. Motivation, which can be enhanced by relevance, meaning, and incentives, is important for successful learning.
4. The degree of association also affects learning.
5. Autonomic arousal may affect the process of learning, although too much arousal can interfere with learning.
6. Older adults take more time to learn and to respond than do younger adults.
7. Some evidence indicates that there is a verbal learning deficit in the later years of life.

IV. Creativity

A. Creativity in Late Adulthood - Outstanding contributions have been made by many people during late adulthood.

B. Creativity as Quality of Production - Lehman found that in most fields people produced the greatest proportion of superior work during their 30s, although valuable contributions were made at all ages.

C. Creativity as Quantity of Work - Dennis, who looked at quantity of work, found that peak performance years occurred later than Lehman maintained, and that it varied depending on the field.

V. Education - Learning is a lifelong endeavor where there is no specific age at which people cease learning.

LEARNING OBJECTIVES/ STUDY QUESTIONS

After reading this chapter, you should be able to:

1. Discuss formal operational thinking and characteristics of thinking for adults.

2. Describe the cognitive style of older adults.

3. Describe the practical problem-solving abilities of older adults.

4. Discuss the abilities of older adults to comprehend word meanings, sentences, and prose.

5. Discuss a possible fifth stage of cognitive development: a problem-finding stage.

6. Discuss dialectical thinking.

7. Summarize Schaie's stages of adult cognitive development.

8. Discuss adult intelligence and the changes in verbal scores and performance scores with age as measured on the WAIS.

9. Describe what happens to scores on the PMA test as people age.

10. Summarize the criticisms of IQ tests.

11. Discuss some of the factors that affect test scores.

12. Describe the relation between personality factors and intellectual functioning.

13. Discuss the relation between socioenvironmental factors and intellectual performance.

14. Describe terminal decline and evaluate the theory.

15. Describe the basic processes involved in memory and the three memory stores.

16. Describe the changes in:
    a. sensory memory -

    b. short-term memory -

    c. long-term memory -

17. Describe the three kinds of long-term memory storage.

18. Discuss extremely long-term memory.

19. Describe the role of memory training in enhancing the memory of elderly adults.

20. Discuss the importance of motivation, relevance, meaning, associative strength, autonomic arousal, and pacing and speed in learning ability.

21. Describe the changes in learning ability with age.

22. Discuss creativity as quality and quantity of production and the changes that take place in creativity during adulthood.

KEY TERMS

In your words, provide a definition for each of the following terms:

1. Field-independent _____

2. Field-dependent _____

3. Dialectical thinking _____

4. Thesis _____

5. Antithesis _____

6. Terminal decline _____

7. Prospective memory _____

8. Retrospective memory _____

9. Acquisition _____

10. Storage _____

11. Retrieval _____

12. Echoic memory _____

13. Iconic memory _____

14. Tactual memory _____

15. Primary memory _____

16. Secondary memory _____

17. Procedural memory _____

18. Semantic memory _____

19. Paired-associate learning _____

20. Serial-learning tasks _____

21. Divided attention tasks _____

22. Intrinsic motivation _____

23. Extrinsic motivation _____

24. Associative strength _____

25. Autonomic arousal _____

26. Pacing _____

APPLICATIONS

For each of the following, fill in the blank with one of the terms listed above.

1. "Dog" and "bone" are related to one another, therefore they have high _____.

2. A memory of a song is part of _____ memory.

3. A person who is given a problem to solve and is able to pick out the relevant information for solving the problem is _____.

4. Janet sees a string tied around her finger and remembers that she is supposed to pick up some groceries on her way home. She is engaging in _____.

5. A person who tries hard at a task because he knows that he will get more money if he does well is working hard because of an _____.

6. Rochelle had to remember to bring in three paper clips to school the next day. This is an example of _____ memory.

7. A person who believes in one opinion but is able to argue for the opposite point of view is able to engage in _____.

8. A task in which a person is expected to learn a list of words in the same order that it is given is a _____.

9. Memory researchers who believe that time of death can be predicted to some degree by a marked decline in intellectual performance believe in the theory of _____.

10. A person who has rehearsed and learned a new telephone number has put that number into _____.

11. The memory of a person's face is part of the _____.

12. Memory of how to cook a grilled cheese sandwich is an example of _____ memory.

SELF-TEST MULTIPLE CHOICE QUESTIONS

Circle the best answer for each question.

1. A person who is given a problem that includes both information that is necessary for solving the problem and irrelevant information and who has trouble determining what is relevant is considered to be
   a. impulsive.
   b. reflective.
   c. field-dependent.
   d. field-independent.

2. Older adults, in contrast to younger adults, tend to be
   a. better at solving problems that involve abstraction.
   b. better at formal reasoning problems.
   c. better at solving practical problems.
   d. all of the above

3. The ability to comprehend complex sentences
   a. remains stable until the 40s, after which it declines.
   b. remains stable until the 60s, after which it declines.
   c. continues to improve throughout the lifespan.
   d. deteriorates slowly from early adulthood onward.

4. Some investigators have suggested that there is a stage beyond formal operations that involves
   a. the ability to discover new problems and raise questions about ill-defined problems.
   b. the ability to engage in abstract thinking.
   c. the ability to reason in a logical and systematic fashion.
   d. the ability to remember details from long ago.

5. Kim believes that capital punishment is morally wrong, yet she is able to consider that others disagree and she can present their side of the argument. Kim is able to engage in
   a. problem-solving.
   b. rational thinking.
   c. problem-finding.
   d. dialectical thinking.

6. According to Schaie, which is the final stage of acquiring and using knowledge?
   a. reintegrative stage
   b. responsibility stage
   c. executive stage
   d. achieving stage

7. In general, verbal scores on the WAIS tend to _____ with increasing age, and performance scores tend to _____.
   a. increase; decrease
   b. decrease; increase
   c. remain the same; decrease
   d. decrease; remain the same

8. Horn and Cattell found that _____ intelligence declined with age, while _____ intelligence increased through adulthood.
   a. verbal; performance
   b. performance; comprehension
   c. crystallized; fluid
   d. fluid; crystallized

9. Research studies comparing IQ scores at different ages may be misleading because
   a. IQ scores are dependent upon which age group the individual is in and are adjusted accordingly.
   b. IQ tests measure innate capacity which would not change over time.
   c. tests are designed for older rather than younger adults.
   d. variability is much greater among older individuals than among younger individuals.

10. Which of the following factors would be most likely to negatively affect test scores?
    a. high degree of motivation
    b. being physically fit
    c. having a dogmatic personality
    d. being self-confident

11. The statement that rigidity increases with age, thus affecting cognitive performance, is too simplistic because
    a. no one is rigid in all aspects of their life, thus it is less likely that rigidity would globally affect cognitive performance.
    b. older people may appear more rigid because they have more trouble shifting from one task to another quickly.
    c. neither of the above
    d. both of the above

12. Which of the following individuals is the least likely to show a decline in cognitive abilities late in life?
    a. an individual who lives in an intellectually stimulating environment
    b. an individual with low socioeconomic status
    c. an individual with very little advanced education
    d. an individual with many life crises

13. Which of the following has been shown to be affected by training of older adults?
    a. crystallized intelligence
    b. fluid intelligence
    c. performance tests
    d. all of the above

14. According to the theory of terminal decline,
    a. a steady decline in cognitive abilities causes death.
    b. any given individual will decline in cognitive abilities only to a certain point and then deterioration will stop.
    c. whatever factors cause death also cause marked decline in cognitive abilities prior to death.
    d. individuals mentally deteriorate at a constant rate starting in the mid-60s.

15. Which of the following is not considered to be a basic process of memory?
    a. acquisition
    b. intelligence
    c. storage
    d. retrieval

16. _____ has a limited capacity, while _____ is basically infinite in its capacity.
    a. Iconic memory; echoic memory
    b. Echoic memory; iconic memory
    c. Long-term memory; short-term memory
    d. Short-term memory; long-term memory

17. Which of the following sensory memory stores shows the fastest decline with increasing age?
    a. visual
    b. auditory
    c. tactual
    d. olfactory

18. Which of the following is an example of primary memory?
    a. being able to remember the positioning of furniture in your house when you are somewhere else
    b. looking at a grocery list and trying to remember the first few items
    c. the ability to remember the faces of your immediate family members
    d. the ability to remember events that happened in your childhood

19. Which of the following tasks would an older individual have the most difficulty with, relative to a younger individual?
    a. a task in which they are shown pictures of different people, and then later, are shown these same pictures and some others and asked if they have seen the people before
    b. a task in which they are given a list of words of different types of food and are asked to memorize them
    c. a task in which they are given a list of unrelated words, and then later asked to recall them
    d. a task in which they are asked to remember items from their own homes

20. One study found that older adults' use of neighborhood resources was related to
    a. spatial cognitive ability.
    b. primary memory.
    c. echoic memory.
    d. semantic memory.

21. Elderly adults are more likely to have problems with _____ memory than with _____ memory.
    a. semantic; procedural
    b. semantic; episodic
    c. procedural; semantic
    d. episodic; procedural

22. A person who is 60 years old is most likely to remember sociohistoric events from when they were _____ years old; a person who is 40 is most likely to remember events from when they were _____ years old.
    a. 30 to 40; 10 to 15
    b. 50 to 60; 30 to 40
    c. 15 to 25; 30 to 40
    d. 15 to 25; 15 to 25

23. Which of the following is an example of a paired-associate learning task?
    a. learning a list of pairs of previously unassociated words, and then being given one term from each pair and asked to remember the other one
    b. learning a list of words and recalling them in the same order
    c. remembering a list of words by associating each one with a word that is related
    d. learning a list of pairs of words that are related, and then being asked to recall both of them in the same order that they were originally given in

24. A person who is driven to learn because she wants to be successful at her chosen career is
    a. intrinsically motivated.
    b. extrinsically motivated.
    c. superficially motivated.
    d. relevantly motivated.

25. If a task is very complex, the level of autonomic arousal for elderly individuals is likely to be
    a. very high due to greater involvement, and thus they will perform better.
    b. very high due to anxiety, and thus they will perform more poorly.
    c. very low due to lack of involvement, and thus they will perform more poorly.
    d. very low due to lack of anxiety, and thus they will perform better.

26. Older adults are most likely to learn paired associates under which of the following conditions?
    a. if the words are presented 4 seconds apart
    b. if the words are sometimes presented 2 seconds apart and sometimes presented 6 seconds apart
    c. if they are allowed to go at their own pace
    d. if more time is given between learning the words and recalling them

27. According to Lehman, individuals in most fields produced their best works during their
    a. 20s
    b. 30s
    c. 40s
    d. 50s

ANSWER KEY

APPLICATIONS

1. associative strength
2. echoic
3. field-independent
4. retrieval
5. extrinsic motivation
6. prospective
7. dialectical thinking
8. serial-learning task
9. terminal decline
10. secondary memory
11. iconic memory
12. procedural

MULTIPLE CHOICE

1. c (p. 530)
2. c (p. 530)
3. b (p. 532)
4. a (p. 533)
5. d (p. 533)
6. a (p. 534)
7. c (p. 534)
8. d (p. 538)
9. a (p. 539)
10. c (p. 541)
11. d (p. 543)
12. a (p. 544)
13. d (p. 545)
14. c (p. 545)
15. b (p. 546)
16. d (p. 547)
17. c (p. 548)
18. b (p. 548)
19. c (p. 549)
20. a (p. 551)
21. d (p. 551)
22. d (p. 553)
23. a (p. 554)
24. a (p. 554)
25. b (p. 556)
26. c (p. 556)
27. b (p. 557)

Chapter 18

EMOTIONAL DEVELOPMENT

CHAPTER OUTLINE

I. Subjective Well-Being - Subjective well-being can best be measured by indicators such as life satisfaction, morale, happiness, congruence, and affect.

   A. Life Satisfaction
      1. Life Satisfaction can be considered to have five dimensions: zest vs. apathy, resolution and fortitude, congruence, self-concept and mood tone.
      2. One study found that the continued fulfillment of meaningful social roles is an important determinant of life satisfaction among African-American elderly.

   B. Sociodemographic Factors
      1. Negative effects of various physical impairments are stronger among blacks than whites.
      2. Satisfaction with one's financial status is more important in determining overall satisfaction than is SES by itself.
      3. Urbanism influenced life satisfaction indirectly and interactively because it influenced health, financial satisfaction, and social integration, but overall, the effects of urban living on life satisfaction are inconsistent.
      4. Minority groups that move to the U.S. have special problems because of language barriers.

   C. Family Relationships - Marital happiness is an important contributor to life satisfaction. Early relationships with parents also affect later well-being.

   D. Morale - The PGC Morale Scale evaluates three aspects of morale: agitation, dissatisfaction, and attitudes toward aging.

   E. Happiness - The young adult years are reported to be the happiest, but the present is reported to be a satisfying time of life as well.

F. Congruence - The *Life Attitude Profile (LAP)* measures seven dimensions of meaning and purpose in life: life purpose, existential vacuum, life control, death acceptance, will to meaning, goal seeking, and future meaning.

G. Personal Goals of Older Adults - There are four types of goals: achievement goals, maintenance goals, disengagement goals, and compensation goals.

H. Affect - Affect, either positive or negative, has been found to be an important dimension of subjective well-being, and has both an hereditary and environmental base.

I. Social Networks - Subjective well-being has also been found to relate to the quality of social networks. Research has found that whether older persons had enough social ties in the objective sense was less important than whether they perceived they had enough.

J. Health - Health is one of the most important factors related to subjective well-being.

K. Stability of Subjective Well-Being - Research has found evidence of the stability of personality and mean levels of psychological well-being in adulthood.

II. Stress

　A. Meaning
　　1. Stress is physical, mental, or emotional strain or tension caused by environmental, situational, or personal pressures and demands.
　　2. Individual differences in how people respond to stress is partially dependent on hereditary factors.
　　3. The amount of stress experienced depends not only on the severity and duration of exposure, but also on one's previous conditioning.

　B. Causes
　　1. Some stress is job-related. When the results of one's work are uncertain or if there are financial problems due to unemployment, stress is increased. Retirement is also a major source of stress.
　　2. Stress can result from role strain, when people feel that they can't manage their situations and feel in control.
　　3. Stress is likely to arise out of interpersonal relationships that are unpleasant and conflicting over a period of time.
　　4. Any kind of transition or change is also stressful for some people.
　　5. Life crises, which are drastic changes in the course of events, can cause stress.
　　6. A great deal of stress is self-induced.

C. Effects of Stress
1. The body goes through three stages in adapting to stress: an alarm reaction, a resistance stage, and then exhaustion.
2. Repeated stress can result in physical damage, decline in health, an increase in illness, and can interfere with psychological functioning.

D. Coping with Stress
1. In a task-oriented approach, adults make a direct effort to alleviate the source of the stress, which requires being aware of what is going on.
2. Another approach to stress is a deliberate, cognitive effort to change one's internal responses.
3. Relaxation theory is used widely to cope with stress.
4. Physical activity and exercise are useful for relieving stress.
5. Another way to deal with stress is to sublimate it through indirect means.
6. A common way of dealing with stress is to take medication.
7. Increasing one's social interests and social supports can minimize stress.
8. Hostile reactions to stress may or may not be helpful.
9. Older adults tend to use more positive coping mechanisms than do adolescents or young adults.

III. Patterns of Family Adjustment to Crises

A. First Stage: Definition and Acceptance
1. The first stage is the onset of the crisis and the increasing realization that a crisis has occurred.
2. The first step involves defining the problem and gradually accepting that a crisis exists.

B. Second Stage: Disorganization - During this period, the family's normal functioning is disrupted. Shock and disbelief may make it impossible to function at all or to think clearly.

C. Third Stage: Reorganization - The third stage is one of gradual reorganization during which family members try to take remedial action.

IV. Mental Illness

A. Depression
1. Depression is the most common functional disorder of adults of all ages.
2. To be diagnosed as suffering from major depression, one must evidence depressed mood in addition to a number of other factors.
3. Major depression can occur in a single episode or alternate with manic phases.
4. No single theory explains the cause of depression. Some research suggests that depression has a hereditary base.

5. Another theory is that depression is related to medical conditions.
6. Learning theory emphasizes that depressed behavior is learned.
7. Cognitive theorists suggest that the way persons interpret a situation is related to depression.
8. Psychodynamic theorists emphasize overreactions to events based on early childhood experiences. Humanistic-existential theorists emphasize that it is caused by excessive strain, worry, and the loss of self-esteem.
9. Some factors correlated with depression in the elderly are: being female, lower income, physical or cognitive impairments, lack of social support, poor health, and pain.
10. An effective and innovative form of treatment for depression is light therapy.

B. Organic Brain Syndromes
1. Organic brain syndromes (OBS) are mental disorders caused by or associated with impairment of brain tissue function.
2. One syndrome is dementia, commonly called senile dementia in older people.
3. OBS is often accompanied by impaired intellectual functioning and emotional symptoms and other disorders.
4. The reversible form of organic brain syndrome is called reversible brain syndrome or acute brain syndrome.
5. Sometimes the disorders are called chronic organic brain syndrome because of permanent brain damage. The two major types are senile psychosis and psychosis associated with cerebral arteriosclerosis. A related cause of senile dementia is a condition called multi-infarct dementia.

C. Alzheimer's Disease - Alzheimer's disease is characterized by rapid deterioration and atrophy of the cerebral cortex, which can afflict people who are in their 40s and 50s or older. Some psychosocial effects may be: restricted life style, discrediting of self, social isolation, feelings of uselessness and of being a burden.

LEARNING OBJECTIVES/ STUDY QUESTIONS

After reading this chapter, you should be able to:

1. Define subjective well-being and describe the specific indicators.

2. Discuss life satisfaction, its meaning, and the socioeconomic factors that correlate with it.

3. Describe morale as a component of life satisfaction.

4. Discuss happiness as a dimension of subjective well-being and discuss what research has found about happiness.

5. Define congruence and discuss its relation to subjective well-being.

6. Describe the four types of personality goals of older adults.

7. Discuss the importance of both social networks and health as components of subjective well-being.

8. Discuss the meaning of stress and its causes and effects.

9. Describe some of the different ways in which adults cope with stress.

10. Describe the three stages of adjustment to family crises.
    a.

    b.

    c.

11. Describe the symptoms of depression.

12. Discuss some of the causes and theories of depression.

13. Discuss the various types of organic brain syndromes.

14. Describe Alzheimer's disease. What are some of the psychosocial effects?

KEY TERMS I

In your own words, provide a definition for each of the following terms:

1. Subjective well-being _____

2. Morale _____

3. Happiness _____

4. Congruence _____

5. Affect _____

6. Social networks _____

7. Stress _____

8. Type A personality _____

9. Crisis _____

10. Crisis overload _____

11. Cognitive modification program _____

12. Relaxation training _____

13. Transcendental meditation _____
_____

APPLICATIONS I

For each of the following, fill in the blank with one of the terms listed above.

1. Kim is very competitive, to the point of seeming hostile to her co-workers. She is always in a rush, and she reacts very strongly to stress. Kim can be said to have a _____.

2. Losing a friend can be considered to be a life _____ because it is a change in a person's life and a turning point that will affect the future.

3. A person who is stressed can use _____ in order to relax the body.

4. Marcus, who has always wanted to be a doctor because he wants to help people, is currently attending medical school. His life, in terms of career goals, can be said to be high in _____.

5. Mary Jane looks back upon her life and is very satisfied with what she has accomplished in terms of her career and her relationships with other people. Mary Jane possesses positive _____.

6. Peter, who is under considerable emotional strain both at work and at home, is experiencing a lot of _____.

7. Gary is in a program in which he is taught different ways of restructuring how he thinks about things, with an emphasis on positive thinking, in order to cope with the stress in his life. This type of program is a _____.

8. All in one week, Melanie's mother died, her child had to be hospitalized and a very important file was lost at work. Melanie is experiencing a _____.

9. Lenny worries a lot and is agitated by little things. He is dissatisfied with his life and believes that it will only get worse as he gets older. Lenny has low _____.

10. _____ is a technique in which the individual tries to divert consciousness away from present troubling thoughts and toward a state of relaxation.

KEY TERMS II

In your own words, provide a definition for each of the following terms:

1. Depression _____

2. Cyclothymic disorder _____

3. Dysthymic disorder _____

4. Monoamines _____

5. Organic brain syndromes _____

6. Senile dementia _____

7. Reversible brain syndrome _____

8. Chronic organic brain syndrome _____

9. Senile psychosis _____

10. Psychosis associated with cerebral arteriosclerosis _____

11. Multi-infarct dementia _____

12. Alzheimer's disease _____

13. Aphasia _____

14. Agnosia _____

15. Apraxia _____

APPLICATIONS II

For each of the following, fill in the blank with one of the terms listed above.

1. When Lewis was 65 years old, he lost the ability to understand language. Lewis has an _____.

2. An individual who more often experiences depression than normal mood has _____.

3. _____ is caused by the atrophy and degeneration of brain cells and is characterized by a decline in mental functioning.

4. _____ is a dementia caused by a series of small strokes that cut off blood supply to the brain.

5. Mel alternates between being extremely depressed and being overly manic. He probably has _____.

6. Karen has lost the ability to recognize the faces of people that she knows even though she can identify them when she hears their voices. She has an _____.

7. An acute brain syndrome that is treatable is a _____.

8. The hardening and narrowing of the blood vessels in the brain can lead to _____.

9. _____ is characterized by rapid deterioration and atrophy of the cerebral cortex.

10. Suzanne has lost the ability to use her hand to write. She may have some type of _____.

SELF-TEST MULTIPLE CHOICE QUESTIONS

Circle the best answer for each question.

1. Which of the following is not a dimension of life satisfaction according to the *Life Satisfaction Index*?
   a. congruence
   b. self-concept
   c. morale
   d. mood tone

2. Which of the following individuals is likely to feel the most satisfied with life?
   a. a person from a wealthier nation in comparison to someone from a poorer nation
   b. a person who feels he is better off then his closest relatives
   c. a person who has more children
   d. a person who makes a lot of money but wishes she were making more

3. A survey by Chiriboga (1978) found that more people reported being the happiest during which period of their lives?
   a. teens
   b. 20s
   c. 30s
   d. 40s

4. The agreement between a person's desired goals and what has actually been achieved is called
   a. congruence.
   b. life satisfaction.
   c. morale.
   d. happiness.

5. A study using the *Life Attitude Profile* found that women, in comparison to men, were
   a. more likely to have a positive acceptance of death.
   b. more likely to want to achieve new goals.
   c. less likely to want to achieve new goals.
   d. more likely to seek meaning to their existence.

6. Subjective well-being is related to social networks in what way?
   a. The person with the most social ties will be the happiest.
   b. The person who believes that she has enough social ties will be the most satisfied.
   c. The person who has his children nearby will be the most satisfied.
   d. The person who sees family members regularly will be the happiest.

7. Longitudinal research on the stability of subjective well-being suggests that
   a. enduring personality dispositions are not important in the long run.
   b. well-being is very variable for the majority of adults.
   c. many adults are able to adapt to varying circumstances and thus maintain their well-being.
   d. all of the above

8. The ability to deal with stress depends on
   a. one's hereditary makeup.
   b. one's previous conditioning.
   c. the severity and duration of exposure.
   d. all of the above

9. Stress can be kept to a minimum if a person
   a. works in a job in which she is exposed to a lot of uncertainty.
   b. works very long hours.
   c. feels she is in control of her situation.
   d. is dissatisfied with his job.

10. Which of the following could be considered a life crisis?
    a. losing a job
    b. having a pet die
    c. an earthquake
    d. all of the above

11. According to Selye, the first stage in adapting to stress is
    a. an alarm reaction.
    b. a resistance stage.
    c. a protest stage.
    d. exhaustion.

12. Which of the following individuals may be more prone toward developing cancer?
    a. a person who expresses a lot of rage
    b. a person who expresses the anxiety that they feel
    c. a person who has difficulty expressing rage and anxiety
    d. a person who rarely experiences stress

13. Philip is feeling some stress because he just bought a new house and he is not sure he can really afford it. He deals with his stress by getting a second part-time job to tide him over until he feels financially secure. He has used what type of approach for dealing with stress?
    a. a cognitive approach
    b. a behavior modification approach
    c. a denial approach
    d. a task-oriented approach

14. Whenever Gail has to give an oral report at work, she feels a lot of anxiety and believes that she will do a poor job. She tells herself, "I'm no good at this; I shouldn't have this job anyway." Gail would probably benefit the most from which type of approach to coping with stress?
    a. a task-oriented approach
    b. a cognitive modification program
    c. relaxation training
    d. exercise

15. Which of the following is an example of using sublimation as a way of coping with stress?
    a. seeking out friends who will listen to problems
    b. becoming angry and trying to directly alleviate the problem
    c. becoming involved in enjoyable hobbies
    d. distancing yourself from the problem

16. During the first stage of adjustment to a family crisis,
    a. the family must define the problem and accept that it exists.
    b. the family is likely to become very disorganized.
    c. stress is at its maximum.
    d. family morale decreases rapidly.

17. During which period of adjustment to family crises are child and spousal abuse most likely to occur if it is going to occur at all?
    a. first stage
    b. second stage
    c. third stage
    d. fourth stage

18. To resolve a family crisis, the family must undergo a stage of
    a. disorganization.
    b. morale building.
    c. stress relaxation.
    d. reorganization.

19. At the end of a period of family crisis, the family
    a. returns to the same way they were before.
    b. is reorganized at a new level which may or may not be as satisfactory as the old one.
    c. will function at a higher level than before because they have experienced a crisis together.
    d. can never function as effectively as they did before.

20. Which of the following is not considered to be a potential symptom of major depression?
    a. poor appetite
    b. loss of energy
    c. increased sexual drive
    d. excessive guilt

21. A person suffering from depression with melancholia is most likely to feel the worst
    a. while waking up.
    b. at the end of the work day.
    c. upon arriving home after work.
    d. before going to bed.

22. Marjorie sometimes feels very sad and low, has a hard time getting out of bed and feels that she is worthless. During other periods, however, she is extremely active, very talkative, easily distracted and she thinks of herself in very grandiose terms. Which of the following diagnoses would fit Marjorie's symptoms the best?
    a. major depression
    b. hypomania
    c. dysthymic disorder
    d. cyclothymic disorder

23. Which theories of depression emphasize overreactions to events based on early childhood experiences?
    a. learning theories
    b. cognitive theories
    c. psychodynamic theories
    d. humanistic-existential theories

24. Which of the following is not a distinguishing feature of senile dementia?
    a. loss of a sensory system such as auditory functioning
    b. impairment of memory
    c. impairment of comprehension
    d. impairment of judgement

25. What is the single most common cause of dementia in the aged?
    a. organic brain syndrome
    b. Alzheimer's disease
    c. senile psychosis
    d. psychosis associated with cerebral arteriosclerosis

ANSWER KEY

APPLICATIONS I

1. Type A personality
2. crisis
3. relaxation training
4. congruence
5. subjective well-being
6. stress
7. cognitive modification program
8. crisis overload
9. morale
10. Transcendental meditation

APPLICATIONS II

1. aphasia
2. dysthymic disorder
3. Senile psychosis
4. Multi-infarct dementia
5. cyclothymic disorder
6. agnosia
7. reversible brain syndrome
8. psychosis associated with cerebral arteriosclerosis
9. Alzheimer's disease
10. apraxia

MULTIPLE CHOICE

1. c (p. 566)
2. b (p. 567)
3. b (p. 570)
4. a (p. 570)
5. d (p. 571)
6. b (p. 573)
7. c (p. 575)
8. d (p. 576)

9. c (p. 576)
10. d (p. 578)
11. a (p. 579)
12. c (p. 579)
13. d (p. 580)
14. b (p. 580)
15. c (p. 581)
16. a (p. 582)

17. b (p. 583)
18. d (p. 583)
19. b (p. 584)
20. c (p. 584)
21. a (p. 584)
22. d (p. 584)
23. c (p. 585)
24. a (p. 587)
25. b (p. 589)

Chapter 19

SOCIAL DEVELOPMENT

CHAPTER OUTLINE

I. Singlehood

   A. Marital Status and Delay - By age 45 to 54, only 7.4% of males and 5.6% of females have never been married.

   B. Typology of Singles - Stein has developed a typology of singles based on whether their status is stable or temporary, voluntary or involuntary.

   C. Advantages and Disadvantages of Being Single
      1. Some reputed advantages of being single are: greater opportunities for self-development and to meet different people, economic independence, more sexual experience, freedom to control one's life, and more opportunities for career change and development.
      2. Some disadvantages are: loneliness, economic hardship, feeling out of place in many social gatherings, sexual frustration, not having children, and prejudice against singles.

   D. Life-Styles - There are at least six life-style patterns of singles: professional, social, individualistic, activist, passive, and supportive.

   E. Living Arrangements - Some of the different living arrangements of singles are: singles communities, shared living spaces, living with parents, living alone, and nonmarital cohabitation.

   F. Friendships and Social Life
      1. One of the greatest needs of single people is to develop interpersonal relationships, networks of friendships that provide emotional fulfillment, companionship and intimacy.
      2. Loneliness is a problem for a significant minority of never-marrieds.
      3. Singles have more time for optional activities.
      4. A greater percentage of married than never-married individuals report getting

more fun out of life and that they were happy.

G. Dating and Courtship
1. The most important element in attraction - at least in initial encounters - is physical attractiveness. Extroversion is considered attractive early on, and later, agreeableness and conscientiousness are important.
2. One of the major problems of those who want to date is where and how to meet prospective partners.
3. Male-female traditional roles in the dating process are changing rapidly.
4. Date rape is a very common occurrence.

H. Sexual Behavior - One study has found that singles are less likely to report being satisfied with their sex lives than are married persons.

I. AIDS is a disease that can affect heterosexuals and homosexuals. It is not transmitted through casual contact.

J. Love and Intimacy
1. Love can be thought of as including these five elements: romantic love, erotic love, dependent love, friendship love, and altruistic love.
2. Sternberg described three components of close relationships: intimacy, passion, and commitment. The most complete love, consummate love, is a combination of all three.

K. Mate Selection - Some factors involved in mate selection are: propinquity, attraction, homogamy and heterogamy, and compatibility.

L. The Older, Never-Married Adult - The major difference between younger and older adults who have never married is that most younger singles consider their status temporary, whereas older singles are often well-adjusted to their situation.

II. Marriage and Family Living

A. Marriage and Personal Happiness - Marital happiness contributes more to personal global happiness than does any other kind of satisfaction, including satisfaction from work.

B. The Family Life Cycle
1. The family life cycle divides the family experience into phases or stages over the life span and seeks to describe changes in family structure and composition during each stage.
2. The general trend is for marital satisfaction to be somewhat curvilinear - to be high at the time of marriage, lowest during the child-rearing years, and higher again after the youngest child has passed beyond the teens.

C. Adjustments Early in Marriage - All couples are faced with marital adjustment tasks in which they try to modify their behavior and relationship to achieve the greatest degree of satisfaction.

D. Adjustment to Parenthood - The more stressful a couple's marriage before parenthood, the more likely it is that they will have difficulty in adjusting to the first child. Stress will vary depending on whether the child was planned and on the child's temperament.

E. Voluntary Childlessness - Couples who choose to remain childless tend to be less traditional and sexist in their views of women.

F. Adjustments During Middle Adulthood
   1. A growing awareness that years are numbered creates a sense of urgency for many middle-agers.
   2. For some, middle age can become a time for revitalizing a tired marriage, for rethinking their relationship, and for deciding that they want to share their life together.
   3. Some couples have trouble adjusting to the empty-nest years, but many are relieved and excited when the last child leaves. These years may be happier than earlier and later years.

G. Spouse Abuse - Spouse abusers tend to have poor self-images, to be aggressive, and they may have alcohol problems. Some may be misogynists. Abused women tend to have very low self-confidence and self-esteem.

H. Adjustments During Late Adulthood
   1. As health and longevity of the elderly increases, an increasing proportion of elderly adults are still living with their spouse.
   2. For many, marital happiness and satisfaction increase during a second honeymoon stage after the children have left home and after retirement.
   3. Sometimes during the last stages of old age, marital satisfaction again declines.
   4. There is some evidence that there is a reversal of sex roles in relation to authority in the family as people get older.
   5. Most older people are not isolated from their adult children.

I. Widowhood - The greater longevity of women means that the number of widows exceeds widowers at all levels. It is helpful to be able to maintain intimate relationships with friends.

J. Divorce
   1. Many unhappy marriages do not end in divorce. Some factors involved are: age, commitment to marriage as an institution, lack of control, and low social activity.

2. Divorce rates increased steadily from 1958 to 1979, and have since leveled off and even declined.
3. Marriage and family therapists have identified 10 areas as having the most damaging effect on marital relationships, with communication problems ranked as the most damaging.
4. The decline of intimacy and love, called disaffection, is a major component in the divorce process.

K. Alternatives to Divorce - Before divorcing, couples may consider marriage counseling, marriage enrichment programs, or separation.

L. Adult Adjustments After Divorce - Some of the adjustments to be made after divorce are: getting over the emotional trauma, dealing with society's attitudes, loneliness and social readjustment, finances, realignment of responsibilities, sexual readjustment, contacts with ex-spouse, and kinship interaction.

M. Extramarital Affairs - Affairs have varying effects on married people and their marriages.

N. Remarriage - Adults who remarry after divorce have some real advantages over those married for the first time, but for some, remarriage introduces new complications.

O. Problems of Stepfathers - Stepfathers have a number of problems with which they have to deal.

P. Problems of Mothers in Remarriages - The mother in a remarriage has a whole set of problems that are unique to her situation.

III. Work and Careers

A. Career Establishment
1. Two major psychosocial tasks of early adulthood are to mold an identity and to choose and consolidate a career.
2. Discontentment with jobs or money affects other aspects of life.
3. Adults have been grouped into five categories depending on the status of their vocational development: vocational achievers, vocationally frustrated, non-committed, vocational opportunists, and social dropouts.

B. Women's Careers
1. Working wives who are able to integrate their home life and work life report greater life satisfaction. Sources of stress include: children, the job itself, and too many roles.
2. Dual careers can create more stress, but can also be very rewarding.
3. Maternal employment will not necessarily affect parent-child relationships.

However, mothers' dissatisfaction with her role, either employed or not, can affect the family negatively.

C. Mid-Life Careers and Employment
 1. For most people, middle adulthood is the fruition of a long period of professional work.
 2. When a person is shut out of a career or at a dead end, one answer is to begin a second career.
 3. Although the middle years can be productive and rewarding for some, for others, they can be years of upset and anxiety.

D. The Older Worker
 1. Most research has found that job satisfaction tends to increase with the worker's age.
 2. The numbers of older workers in the work force may increase in the years ahead.
 3. Forced retirement has been ranked among the top 10 crises in terms of the amount of stress it causes the individual.

IV. Social-Psychological Theories of Aging

A. Disengagement Theory
 1. This theory states that as people approach old age, they have a natural tendency to withdraw socially and psychologically from the environment, thereby freeing them from various responsibilities, allowing them more time and enhancing their life satisfaction.
 2. Critics argue that disengagement is not universal, inevitable, or inherent in the aging process, nor does it consider individual differences in health and personality.

B. Activity Theory
 1. This theory suggests that an active life-style will have a positive effect.
 2. Critics argue that the theory is oversimplified and does not consider those who cannot be active.

C. Personality and Life-Style Theory - In order to study individual differences, some gerontologists have considered different personality types, such as: integrated, armored-defended, passive-dependent, and unintegrated.

D. Exchange Theory - According to this theory, older individuals who are dependent will lose power to those who are not as needy.

E. Social Reconstruction Theory - This theory states that our society brings about role loss for the elderly, then labels them negatively and deprives them of opportunities, leading to acceptance of the external labeling.

LEARNING OBJECTIVES/ STUDY QUESTIONS

After reading this chapter, you should be able to:

1. Discuss the incidence of singlehood and the 4 categories of singles.

2. Discuss both the advantages and the disadvantages of being single.

3. Describe the life style patterns and living arrangements of singles.

4. Discuss friendships and social life, dating and courtship, and sexual behavior of singles.

5. Discuss the five different types of love.
a)

b)

c)

d)

e)

6. Describe the eight combinations of Sternberg's three components of love.

7. Discuss the situations of older, never-married adults.

8. Discuss the relation between marriage and personal happiness.

9. Outline the stages of the family life cycle.

10. Summarize common adjustments early in marriage.

11. Discuss the adjustments to parenthood.

12. Describe the marital adjustments during middle adulthood.

13. Discuss the characteristics of spouse abusers.

14. Describe the marital adjustments during late adulthood.

15. Discuss the incidence of divorce, the problems that lead to divorce, and the primary adult adjustments after divorce.

16. What are the alternatives that couples sometimes choose before divorce?

17. Discuss extramarital affairs and their varying effects on marriage.

18. Describe some advantages and problems of remarriage.

19. Summarize the process of career establishment.

20. Discuss some of the issues involved for women choosing careers.

21. Discuss mid-life careers and employment.

22. Discuss issues in relation to the older worker and retirement.

23. Summarize the social-psychological theories of aging and evaluate them.
    a. Disengagement theory -

    b. Activity theory -

c. Personality and life-style theory -

d. Exchange theory -

e. Social reconstruction theory -

KEY TERMS

In your own words, provide a definition for each of the following terms:

1. Cohabitation_____

2. Date rape_____

3. Romantic love_____

4. Erotic love_____

5. Dependent love_____

6. Friendship love_____

7. Altruistic love_____

8. Consummate love_____

9. Homogamy_____

10. Heterogamy_____

11. Family life cycle_____

12. Marital adjustment tasks_____

13. Postparental years_____

14. Misogynist _____

15. Marriage enrichment programs _____

16. Structured separation _____

17. Burnout _____

18. Disengagement theory _____

19. Activity theory _____

20. Personality and life-style theory _____

21. Exchange theory _____

22. Social reconstruction theory _____

APPLICATIONS

For each of the following, fill in the blank with one of the terms listed above.

1. According to _____ theory, the people that are most dependent lose the most power, while those who meet those needs gain the power.

2. According to _____ theory, aging people naturally withdraw from society, and in so doing, they free themselves of many responsibilities.

3. The idea that society affects the self-esteem of the elderly by having certain expectations is part of the _____ theory.

4. Diane and Tom live together with two children and are not married. Their living arrangement is that of _____.

5. When Cheryl's and Earl's last child leaves for college, they will be entering the _____.

6. Paul distrusts most women. He is always suspicious of his girlfriend and he tries to control her every move. He blames her and women in general for his own inability to be successful. Paul is a _____.

7. Ruth married a man who was very similar to her in terms of background, habits, interests, and likes and dislikes. This is an example of _____.

8. Ruth and her husband enjoy participating in the same activities and in general, enjoy being with one another. They have _____ love.

9. Barry has been working at his job for many years and recently has felt emotionally and physically exhausted from the pressure. He is suffering from _____.

10. Jack and Joanne go out on a date and return to her apartment afterwards. They kiss and pet for a while, and then Jack forces Joanne to have intercourse even though Joanne has said that she doesn't want to. This is an example of a _____.

11. The _____ describes the stages and changes in family structure, and the challenges and tasks that the family faces over the life span.

12. The theory which states that an elderly person who continues to be active will have a more positive outlook on life is the _____ theory.

13. The theory that considers the relation between personality types, such as passive-dependent types, and patterns of aging is the _____ theory.

14. Mary Jane and Lou have emotionally bonded, are passionate in their love-making, and are committed to each other. According to Sternberg, they have _____ love.

15. Lateka is deeply concerned about her husband. She enjoys caring for him and giving to him. She feels _____ love.

16. Timothy is extremely attracted to Vicki. He feels _____ love toward Vicki.

SELF-TEST MULTIPLE CHOICE QUESTIONS

Circle the best answer for each question.

1. By age 45 to 54, about ____ of males and females have never been married.
   a. 5-7%
   b. 15-20%
   c. 20-25%
   d. 30-33%

2. Maria has never been married, but she would like to be and is actively looking for a mate. She would be classified as a
   a. voluntary temporary single.
   b. voluntary stable (permanent) single.
   c. involuntary temporary single.
   d. involuntary stable (permanent) single.

3. Fred is single and a doctor. His major concern in life is his work, and he organizes his social life around his work. Fred would be classified as having what type of single life-style pattern?
   a. professional
   b. social
   c. individualistic
   d. passive

4. Shari is single and not much interested in pursuing an active social life. Instead, she enjoys her hobbies and spends a lot of time alone, enjoying her freedom and doing the things she likes to do best. She would be classified as having what type of single life-style pattern?
   a. supportive
   b. passive
   c. activist
   d. individualistic

5. Who of the following individuals is more likely to be living alone?
   a. a 22 year old woman
   b. a 22 year old man
   c. a 30 year old woman
   d. a 30 year old man

6. Research looking at the effects on premarital cohabitation has found that couples who have lived together
   a. are no more likely than couples who have not lived together before marriage to have successful marriages.
   b. for more than 2 years are more likely to have successful marriages than those who have not.
   c. for less than 2 years are most likely to have successful marriages.
   d. for more than 2 years are less likely to have successful marriages than those who have not.

7. Research suggests that the most important factor in attraction in initial encounters is
   a. extroversion.
   b. agreeableness.
   c. physical attractiveness.
   d. emotional stability.

8. Romantic love
    a. is the best indication that a couple is suitable for one another.
    b. involves strong physical attraction.
    c. is characterized by intense emotions, tenderness, and affection.
    d. all of the above

9. According to Sternberg, fatuous or foolish love is missing which component(s)?
    a. passion
    b. intimacy
    c. commitment
    d. intimacy and commitment

10. The major difference between younger and older adults who have never married is that
    a. most younger adults are happy with their situation, whereas the older adults tend to be more dissatisfied.
    b. most younger singles consider their status temporary, whereas older adults tend to be more adjusted to their status.
    c. younger adults tend to be more sexually active than older singles.
    d. older singles tend to be happier with their situation than do younger singles.

11. In general, marital satisfaction tends to be lowest
    a. within the first few months of marriage.
    b. before having children.
    c. during the child-bearing years.
    d. after the last child leaves home.

12. Adjustment to parenthood can be particularly difficult
    a. if the pregnancy was not planned.
    b. because the transition is quite abrupt.
    c. if the child has a difficult temperament.
    d. all of the above

13. Research suggests that adults in the postparental period tend to be
    a. extremely depressed because their children have left.
    b. confused as to their new roles.
    c. happier than those that are younger or older.
    d. more unhappy with their marriages.

14. Which of the following is often characteristic of spouse abusers?
    a. high self-esteem
    b. high in general aggression
    c. high self-confidence
    d. none of the above

15. Divorce rates are
    a. rapidly declining.
    b. steadily increasing.
    c. leveling off.
    d. constantly up and down.

16. What factor have adults rated as the most damaging to marriages?
    a. lack of communication
    b. unrealistic expectations of marriage or spouse
    c. power struggles
    d. serious individual problems

17. During which phase of the process of marital disaffection are the partners optimistic about their future marriage?
    a. beginning
    b. middle
    c. end
    d. none of the above

18. Marriage enrichment programs are best entered
    a. when one spouse wants to participate but the other one does not.
    b. before problems have become unmanageable.
    c. when the relationship has deteriorated to a point where it is difficult for them to solve problems.
    d. if the couple wants to reunite after a divorce.

19. Which of the following is not an advantage that adults who remarry after divorce have over adults marrying for the first time?
    a. They tend to be older and more mature.
    b. They tend to be more motivated to make their marriages work.
    c. They believe that they are better able to communicate.
    d. There are more likely to be children involved in the remarriage.

20. Rebecca has just completed her M.B.A. and is looking forward to her new job with a prestigious advertising agency. She would be classified as
    a. a vocational achiever.
    b. vocationally frustrated.
    c. noncommitted.
    d. a vocational opportunist.

21. Jacob isn't really sure what he wants to do for a career. He's had a number of different jobs, and now he has taken a job at his uncle's company until he can find something better. He would be classified as
    a. vocationally frustrated.
    b. a vocational opportunist.
    c. noncommitted.
    d. a social dropout.

22. The most successful dual-career families are those in which
    a. husband and wife are treated as equal partners.
    b. neither partner needs to commute long distances.
    c. the couple is happy with child-care arrangements.
    d. all of the above

23. Whether or not retirement causes a lot of stress depends primarily upon
    a. whether their adult children are willing to support them.
    b. whether they had a choice.
    c. whether they felt they had performed well at their job.
    d. whether they have a second career, even if part-time, in mind.

24. Which of the following theories suggests that as people enter old age, there is a natural tendency to withdraw socially and psychologically?
    a. disengagement theory
    b. activity theory
    c. personality and life-style theory
    d. exchange theory

25. Which of the following theories is the most focused toward individual differences?
    a. disengagement theory
    b. personality and life-style theory
    c. exchange theory
    d. social reconstruction theory

26. Marty is afraid of growing old so he works very hard in order to maintain a life-style such as he had when he was a little bit younger. He believes that this will keep him from aging. Marty would most likely be classified as having which personality type?
    a. integrated: reorganized
    b. integrated: focused
    c. armored-defended: holding on
    d. armored-defended: constricted

27. In his older years, Will has become very dependent, but he is basically satisfied because his younger son attends to his needs. Which personality type is Will?
    a. armored-defended: constricted
    b. unintegrated
    c. passive-dependent: succorance-seeking
    d. passive-dependent: apathetic

28. Which theory deals with issues of power in relation to dependency needs?
    a. disengagement theory
    b. activity theory
    c. social reconstruction theory
    d. exchange theory

29. According to social reconstruction theory, negative changes in the self-concept are caused by
    a. society's expectations and labels.
    b. mistreatment by family members.
    c. apathy about their life due to inactivity.
    d. problems with their health.

30. According to social reconstruction theory, people can change the system and break the cycle if
    a. the idea that work is the primary source of self-worth is eliminated.
    b. the social services and health care of the aged are improved.
    c. the elderly are given greater powers of self-rule.
    d. all of the above

ANSWER KEY

APPLICATIONS

1. exchange
2. disengagement
3. social reconstruction
4. cohabitation
5. postparental years
6. misogynist
7. homogamy
8. friendship
9. burnout
10. date rape
11. family life cycle
12. activity
13. personality and life-style
14. consummate
15. altruistic
16. erotic

MULTIPLE CHOICE

1. a (p. 596)	11. c (p. 612)	21. b (p. 628)
2. c (p. 596)	12. d (p. 613)	22. d (p. 629)
3. a (p. 600)	13. c (p. 616)	23. b (p. 632)
4. d (p. 600)	14. b (p. 617)	24. a (p. 632)
5. d (p. 601)	15. c (p. 620)	25. b (p. 633)
6. a (p. 603)	16. a (p. 620)	26. c (p. 633)
7. c (p. 602)	17. a (p. 621)	27. c (p. 633)
8. c (p. 606)	18. b (p. 623)	28. d (p. 634)
9. b (p. 608)	19. d (p. 625)	29. a (p. 635)
10. b (p. 609)	20. a (p. 627)	30. d (p. 635)

Chapter 20

DEATH, DYING, AND BEREAVEMENT

CHAPTER OUTLINE

I. Leading Causes of Death - Heart disease is the number one killer in the U.S., followed by cancer. The causes of death have changed radically in the past 100 years.

II. Attitudes Towards Death and Dying

   A. Cultural Antecedents - In our society, we often hide or deny death, which can sometimes prevent acceptance and positive adjustments.

   B. Criticisms - Currently, there is a swing away from the denial of death toward the notion of death with dignity.

   C. Attitudes Among Different Age Groups of Adults - Middle-aged respondents expressed the greatest fear of death; the elderly expressed the least.

   D. Attitudes Among the Elderly - The elderly are often more philosophical, more realistic, and less anxious about death than are others in our culture.

III. Aspects of Death

   A. What is Death? - A distinction can be made between physiological death, clinical death, sociological death, and psychic death.

   B. Patterns of Death - Pattison outlined 5 patterns of death.

IV. Varying Circumstances of Death

   A. Uncertain Death - The circumstances of uncertain death can be very stressful for the individual and people who are close to him or her.

   B. Certain Death
      1. Kubler-Ross identified five stages of dying that did not necessarily occur in a

regular sequence: denial, anger, bargaining, depression, and acceptance.
2. According to Pattison, in between a period of crisis and the knowledge of death is the living-dying interval, which can be divided into three phases: the acute crisis phase, the chronic living-dying phase, and the terminal phase.
3. Dying people have a number of needs, and usually have a number of tasks to complete.
4. Most people agree that patients have a right to know that they are dying, but how the patients are told must be dealt with on an individual basis.
5. The hospice has emerged as a viable alternative to hospital deaths that are depersonalized, lonely and painful experiences.

C. Anticipatory Grief - a process of emotional detachment while emotional involvement with the dying person is still maintained.

D. Untimely Death: Premature Death
1. The psychological reactions to death are more extreme when death occurs in childhood or at a comparatively young age.
2. Sudden infant death syndrome (SIDS) is the most common cause of death in infants between 2 weeks and 1 year of age.

E. Untimely Death: Unexpected Death - The emotional impact on survivors is gauged by how vital, alive and distinctive the person is at the time of death.

F. Untimely Death: Calamitous Death - Calamitous death is not only unpredictable, but it can be violent, destructive, demeaning, and even degrading.

G. Calamitous Death: Homicide - Most of the stereotypes about murder and murderers have no foundation in fact. Homicide is most often an outgrowth of quarrels and violence among family members or friends.

H. Calamitous Death: Suicide -
1. Suicide rate increases with increasing age. Females attempt suicide more frequently than do males, but more males than females are successful at completing the suicide.
2. Older people who attempt suicide fail much less often than younger people.
3. Some factors that contribute to adolescent suicide are: disturbed family backgrounds, frequent parental absence, social isolation, depression, alcohol and drug abuse, immature personalities, suggestibility, mental illness, guilt, hostility, or attempts for attention and sympathy.

I. Socially Accelerated Dying - In the broadest sense, socially accelerated dying is allowing any condition or action of society that shortens life and hastens death.

V. Euthanasia

    A. Meaning - Euthanasia can be described as a positive/active process (forcing a person to die) or a negative/passive process (doing nothing).

    B. Death with Dignity - Death with dignity allows a terminally ill patient to die naturally without mechanized prolongation that could turn death into an ordeal. Often this is achieved through the use of a living will.

    C. Mercy Killing - Mercy killing is positive, active, direct euthanasia, either voluntary or involuntary.

    D. Death Selection - Opponents of euthanasia are especially concerned about death selection, the involuntary or even mandatory killing of persons who are no longer considered socially useful or who are judged to be a burden on society.

VI. Bereavement

    A. Grief Reactions - Grief reactions are dealt with on four levels: physical, emotional, intellectual, and sociological.

    B. Bereavement in Children - The reaction of a child to grief depends upon age, understanding of death, the reactions of family members, and the child's relationship with the deceased.

    C. The Stages of Grief - There are usually three stages of grief: a short period of shock, a period of intense suffering, and a gradual reawakening of interest in life.

    D. Gender Differences - Men have been conditioned not to show their emotions. Women not only have more friends than men, but women more often use these friends as supports in time of loss.

    E. Cultural Differences - Cultural differences, such as religious beliefs, play a part in how people cope with death.

LEARNING OBJECTIVES/ STUDY QUESTIONS

After reading this chapter, you should be able to:

1. Summarize the leading causes of death today.

2. Discuss the attempts to make death invisible, to deny its fact.

3. Describe the attitudes toward death among middle-aged and older adults.

4. Describe children's and adolescent's conceptions of death.

5. Describe the four aspects of death.
   a.

   b.

   c.

   d.

6. Summarize Pattison's five patterns of death.
   a.

   b.

   c.

   d.

   e.

7. Summarize the varying circumstances of death and the adjustments needed under each circumstance.

8. Identify the five stages of dying as outlined by Kubler-Ross.
   a.

   b.

   c.

   d.

   e.

9. Describe the living-dying trajectory as outlined by Pattison.

10. Discuss the difference between disintegrated and integrated dying.

11. Describe the needs of the terminally ill.

12. Discuss the types of untimely death:
    a. premature death -

b. unexpected death -

c. calamitous death -

13. Discuss death from involuntary manslaughter, homicide, and suicide.

14. Discuss some of the factors that contribute to adolescent suicide.

15. Describe the different types of socially accelerated dying.

16. Discuss grief reactions and the stages of grief.

17. Discuss gender differences in relation to grief.

18. Describe the reactions of children to grief.

KEY TERMS

In your own words, provide a definition for each of the following terms:

1. Physiological death _____

2. Clinical death _____

3. Sociological death _____

4. Psychic death _____

5. Hospice _____

6. Living-dying interval _____

7. Disintegrated dying _____

8. Integrated dying _____

9. Anticipatory grief _____

10. Sudden infant death syndrome _____

11. Posttraumatic stress disorder _____

12. Socially accelerated dying _____

13. Euthanasia _____

14. Death with dignity _____

15. Living will _____

16. Mercy killing _____

17. Death selection _____

18. Idealization _____

APPLICATIONS

For each of the following, fill in the blank with one of the terms listed above.

1. When Marianne checked in on her 2 month old baby, she found him dead in his crib. He had been a healthy baby, and she had no reason to suspect that he was ill. He may have died from _____.

2. After Mark's elderly mother was rushed to the hospital, one of the choices that Mark had to make was whether his mother should be hooked up to various life support machines. Since she had no hope of recovery and she had previously told Mark that she would not want to be kept alive under these conditions, Mark chose not to have any artificial life supports used so that his mother could have _____.

3. After a serious accident, Deborah's brain ceased to exhibit any activity, although other parts of her body could continue to function if aided by life support machines. Deborah could be said to have experienced _____ death but not _____ death.

4. Marjorie made up a _____ in case she was ever in a situation where someone had to decide whether to keep her alive even though there was little or no chance of recovery. In it she specified that she would not want life-sustaining procedures to be done when there was no reasonable expectation of recovery.

5. The act of allowing a person to die naturally without life support or actively putting to death a person who suffers from an incurable disease is called _____.

6. When David learned that he was dying, at first he became very upset. Gradually, however, he began to deal with his impending death and decided to live life to the fullest while he could. Although sometimes he didn't feel well enough to do anything and he would sometimes be depressed, at other times, he was able to continue with his life and often enjoy himself. This is an example of _____.

7. Because dying in a hospital away from family members can be painful, lonely and depressing, many people choose to go to a _____, where they are made to feel as comfortable as possible with their family members present.

8. Marshall's wife Gloria was dying of cancer and was in great pain. She had no hope of recovery, and had expressed to her husband that she no longer wanted to live in this condition. Marshall gave her some pills that killed her, believing this to be the most humane way to deal with the situation and what she wanted. This is an example of _____.

9. When Victoria was 85 years old, she was put into a nursing home where family members visited her occasionally. After a few years, she became terminally ill. Her family members quickly stopped visiting her although she did not die for another few years. Before her physiological death, Victoria could be said to have experienced _____.

10. According to Pattison, the interval between learning about impending death and the death itself is the _____.

11. Any society that allows pollution to escalate in a way that can threaten public health, potentially shortening lives, is contributing to _____.

12. Many soldiers, after having served in a war and seeing many atrocities, experience _____ in which they are extremely emotionally upset.

SELF-TEST MULTIPLE CHOICE QUESTIONS

Circle the best answer for each question.

1. The leading cause of death in the United States is
   a. cancer.
   b. homicide.
   c. heart disease.
   d. AIDS.

2. Which of the following individuals is likely to express the greatest fear of death?
   a. a 16-year-old.
   b. a 25-year-old.
   c. a 50-year-old.
   d. a 70-year-old.

3. A child who believes that death is temporary and reversible is likely to be between the ages of
   a. 3 to 5.
   b. 5 to 9.
   c. 9 to 12.
   d. 13 to 15.

4. A patient who knows that he is going to die and withdraws from the world, regressing into the self, can be said to have experienced a
   a. physiological death.
   b. clinical death.
   c. sociological death.
   d. psychic death.

5. One of the most difficult aspects of dealing with uncertain death is
   a. knowing that someone else created the painful situation.
   b. knowing that the doctors can't help in this situation.
   c. dealing with the guilt.
   d. waiting for the outcome.

6. When Grace was told by her doctor that she had an inoperable brain tumor and would likely die within the next several months, she refused to believe it and thought that the doctors must have made a mistake. According to Kubler-Ross, Grace is at what stage of dying?
   a. denial
   b. anger
   c. bargaining
   d. depression

7. After Grace acknowledged that the doctors were correct about her inoperable brain tumor, she became very hostile and bitter. She began arguing with her family and her health care providers about everything they did. Now Grace could be said to be in the stage of
   a. denial
   b. anger
   c. depression
   d. acceptance

8. Which of the following is not a phase of the living-dying interval?
   a. the acute crisis phase
   b. the chronic living-dying phase
   c. the terminal phase
   d. the acceptance phase

9. Rupert became so depressed after finding out that he was dying that he could no longer deal with daily functioning. He stopped eating and rarely got himself out of bed, even though physically, he was still healthy enough to do so. He lost the will to live, and just let himself become sicker and sicker. According to Pattison, this is an example of
   a. integrated dying.
   b. disintegrated dying.
   c. severe depression accompanying chronic illness.
   d. a phase of disillusionment.

10. Which of the following is generally considered to be the best philosophy when it comes to telling patients they are dying?
    a. A patient should always be told that they are dying even if it would create enormous stress and shock which could harm them.
    b. If a person appears to be leading a happy life and there isn't anything that can be done to save them, they shouldn't be told they are dying because it will only upset them.
    c. In general, every patient has a right to know that they are dying, but each case should be considered individually for when and what they should be told.
    d. It is always better to tell a family member first and let them decide when and how the patient should be told that they are dying.

11. What is the most common cause of death in infants between 2 weeks and 1 year?
    a. chromosomal abnormalities
    b. rubella
    c. sudden infant death syndrome
    d. meningitis

12. Which of the following is a myth about murders or murderers?
    a. Most murderers are not strangers.
    b. Police are more likely to be killed while investigating domestic disturbances than in any other type of duty.
    c. The incidence of violent crime decreases with increases in neighborhood income.
    d. In most cases of violence, the perpetrator and the victim are of different races.

13. Which of the following is the best statement about suicide?
    a. Females and males attempt suicide at the same rate, but males are more likely to actually kill themselves.
    b. Females and males attempt suicide at the same rate, but females are more likely to actually kill themselves.
    c. Females attempt suicide more often than do males, but males are more likely to complete the suicide.
    d. Males attempt suicide more often than do females, but females are more likely to complete the suicide.

14. Are older or younger individuals more likely to complete a suicide attempt?
    a. Older individuals are more likely to complete a suicide attempt.
    b. Younger individuals are more likely to complete a suicide attempt.
    c. Older and younger are equally likely to complete a suicide attempt.
    d. Only older females are more likely to complete a suicide attempt.

15. Which of the following is a myth about adolescent suicide?
    a. Attempted suicide can be a cry for help or attention from other people.
    b. Suicide attempts by adolescents are usually spur of the moment decisions.
    c. Suicidal adolescents tend to come from disturbed family backgrounds.
    d. The risk of suicide among adolescents is increased with alcohol and drug abuse.

16. Which of the following is an example of a condition that can lead to socially accelerated dying?
    a. industrial pollution that is not controlled
    b. unhealthy work conditions
    c. exposure to radiation
    d. all of the above

17. Which of the following allows a terminally ill person to die naturally without putting them on machines that prolong life but may be a terrible ordeal?
    a. death with dignity
    b. mercy killing
    c. death selection
    d. active euthanasia

18. If a person would like to specify that life-sustaining procedures be withdrawn if there is no hope for recovery, they should
    a. tell a family member that this is what they prefer.
    b. tell their family physician that this is what they prefer.
    c. write up a living will.
    d. include this preference in their last will and testament.

19. What is the major difference between mercy killing and death with dignity?
    a. Mercy killing is a type of euthanasia whereas death with dignity is not.
    b. Death with dignity is undertaken to relieve a person of suffering, whereas mercy killing is not.
    c. Death with dignity allows for a natural death, while mercy killing is an active process.
    d. Death with dignity is undertaken by the hospital staff, whereas mercy killing is done by a family member or close friend.

20. Which type of euthanasia causes the most concern for the aged, and for people with severe handicaps or mental retardation?
    a. mercy killing
    b. death selection
    c. death with dignity
    d. eugenics

21. According to a study of grief reactions, the most common psychological grief reaction is
    a. thinking or talking about the patient.
    b. questioning the fairness of death.
    c. an increase in illnesses.
    d. feeling relieved when the ordeal is over with.

22. Which of the following statements made by a person who was grieving is the best example of *idealization?*
    a. From a good friend of the deceased: "It's just as well that he died. He was really suffering."
    b. From a close relative who was very fond of the deceased: "Nobody deserves to die the way he did."
    c. From a close relative who never really got along with the deceased: "It's a shame that his children have to suffer so terribly."
    d. From a close relative who never really got along with the deceased: "He was such a good man, it's such a shame we'll never see him again."

23. If idealization of the deceased continues for a while,
    a. it can help shorten the grieving process.
    b. it can prevent the formation of new intimate relationships.
    c. it can relieve the depression that normally accompanies the grieving process.
    d. it can help lead to a happier new life.

24. According to Hiltz, what occurs during the third stage of the stages of grief?
    a. The bereaved feel as if they are in a state of shock.
    b. The bereaved show physical and emotional symptoms of great disturbance.
    c. There is a painful longing for the dead.
    d. There is a gradual reawakening of interest in life.

25. Which of the following has been found to be a gender difference in how grief is handled?
    a. Widowers are more likely to feel lonely and depressed than are widows.
    b. Widowers are less willing to talk about their feelings associated with their loss than are widows.
    c. Widows are more likely to rely on friends as supports in times of loss than are widowers.
    d. all of the above

ANSWER KEY

APPLICATIONS

1. sudden infant death syndrome
2. death with dignity
3. clinical; physiological
4. living will
5. euthanasia
6. integrated dying
7. hospice
8. mercy killing
9. sociological death
10. living-dying interval
11. socially accelerated dying
12. posttraumatic stress disorder

MULTIPLE CHOICE

1. c (p. 644)
2. c (p. 647)
3. a (p. 647)
4. d (p. 648)
5. d (p. 650)
6. a (p. 651)
7. b (p. 652)
8. d (p. 652)
9. b (p. 652)
10. c (p. 653)
11. c (p. 655)
12. d (p. 656)
13. c (p. 657)
14. a (p. 658)
15. b (p. 659)
16. d (p. 658)
17. a (p. 660)
18. c (p. 660)
19. c (p. 662)
20. b (p. 662)
21. a (p. 663)
22. d (p. 663)
23. b (p. 663)
24. d (p. 663)
25. d (p. 664)